TIME

What is time? St Augustine famously claimed that he knew the answer as long as no one asked him for it, but as soon as he tried to explain it he no longer knew. Part of the problem is the intricate nature of the question. Every individual will approach the question 'What is time?' from a different perspective. We find ourselves asking whether time is linear or cyclic, whether it is endless, whether it is possible to travel in time, how the experience of the flow of time arises, how our own internal clocks are regulated and how our language captures the temporality of our existence.

In this volume nine eminent researchers explore how investigations in their respective fields impinge on questions about the nature of time. These fields encompass the entire range from the arts and humanities to the natural sciences, mirroring the truly interdisciplinary nature of the subject.

THE DARWIN COLLEGE LECTURES

TIME

Edited by *Katinka Ridderbos*

PUBLISHED BY THE PRESS SYNDICATE OF THE UNIVERSITY OF CAMBRIDGE
The Pitt Building, Trumpington Street, Cambridge, United Kingdom

CAMBRIDGE UNIVERSITY PRESS
The Edinburgh Building, Cambridge CB2 2RU, UK
40 West 20th Street, New York, NY 10011-4211, USA
477 Williamstown Road, Port Melbourne, VIC 3207, Australia
Ruiz de Alarcón 13, 28014 Madrid, Spain
Dock House, The Waterfront, Cape Town 8001, South Africa

http://www.cambridge.org

© Darwin College, Cambridge 2002

This book is in copyright. Subject to statutory exception
and to the provisions of relevant collective licensing agreements,
no reproduction of any part may take place without
the written permission of Cambridge University Press.

First published 2002

Printed in the United Kingdom at the University Press, Cambridge

Typeset in 10/14 IronSB QuarkXpress [wv]

A catalogue record for this book is available from the British Library

Library of Congress Cataloguing in Publication data

Time / edited by Katinka Ridderbos.
    p. cm. – (The Darwin College lectures)
  Includes bibliographical references and index.
  ISBN 0 521 78293 7
    1. Time. I. Ridderbos, Katinka, 1970– II. Series.
  QB209 .T48 2002
  115–dc21   2001052875

ISBN 0 521 78293 7 hardback

Contents

	Introduction KATINKA RIDDERBOS	1
1	Time and Modern Physics CHRISTOPHER J. ISHAM AND KONSTANTINA N. SAVVIDOU	6
2	Cyclic and Linear Time in Early India ROMILA THAPAR	27
3	Time Travel D. H. MELLOR	46
4	The Genetics of Time CHARALAMBOS P. KYRIACOU	65
5	The Timing of Action ALAN WING	85
6	Talking about Time DAVID CRYSTAL	105
7	Storytime and its Futures GILLIAN BEER	126
8	Time and Religion J. R. LUCAS	143
	Notes on Contributors	166
	Acknowledgements	169
	Index	171

The plate section is between pp. 26 and 27.

Introduction

KATINKA RIDDERBOS

What is time? We are all too familiar with our inability to escape the inexorable passage of time, and yet we are hard-pressed to say what time is. St Augustine famously replied to this question by saying that he knew the answer as long as no one asked him for it, but that as soon as he tried to explain it he no longer knew. Are we now, more than a millennium and a half after St Augustine found himself in this unfortunate position, better equipped to shed some light on the nature of time?

St Augustine's problem stems at least in part from the fact that the question of the nature of time is multifaceted in a manner in which few other subjects are. Each questioner will emphasise a different aspect of the question 'What is time?'. So, from different perspectives, we find ourselves asking whether time is linear or cyclic, whether it is endless, whether it is possible to travel in time, how the experience of the flow of time arises, how our own internal clocks are regulated and how our language captures the temporality of our existence.

The contributions to this volume reflect this multifacetedness. No attempt is made to devise a single conclusive answer to the question of the nature of time. Instead, eight eminent researchers explore how investigations in their respective fields impinge on questions about the nature of time. These fields encompass the entire range from the arts and humanities to the natural sciences, mirroring the truly interdisciplinary nature of the subject.

Christopher Isham, Professor of Theoretical Physics at Imperial College London, and his co-worker Konstantina Savvidou analyse our understanding of the role that time plays in modern physical theories. Or rather 'roles', since the idea of time arises in two distinct ways in descriptions of the physical world. First, such descriptions comprise statements concerning the way things *are* at any given moment of time; here time is viewed from the perspective of 'being'. But a physical description of the world also requires statements as to how things *change*; here time refers to the complementary notion of 'becoming'. The standard mathematical representation of time in physical theories conflates these two roles. In classical physics this situation does not lead to any immediate problems. However, an exciting possibility to emerge from chapter 1 is that distinguishing between these two roles of time more sharply than is usually the case might be profitable for the future development of theoretical physics. This might in particular be the case for the theory of quantum gravity, in which time plays a notoriously problematic role.

Modern physical theories take for granted that time is linear, with a strict separation of temporal experience into past, present and future. However, a competing notion of time existed for example in early Indian civilisation, according to which time is cyclic. The notion of cyclic time suggests an endless repetition of events, without strongly demarcated points of beginning or end. Such a cyclic world view has often been said to preclude a sense of history. The idea that early Indian civilisation was ahistorical is unravelled by Romila Thapar, Emeritus Professor of History at the Jawaharlal Nehru University in New Delhi, as an idea founded in the preconceptions of European scholars who were keen to 'discover' what they preferred to think of as the uncharted territory of Indian history. Drawing on her extensive knowledge of early Indian texts, Professor Thapar identifies a rich tradition of historical chronologies. The linear concept of time required for these chronologies co-existed with the cyclic notion of time; both were used, often simultaneously, but for different purposes. A pervasive sensitivity to the different functions of each of them enhanced the meaning of both.

In a cyclic universe, each event that lies in the past of the present moment, also lies in its future. Thus no specific effort needs to be made by a time traveller who desires to revisit the past: it suffices to wait for events to repeat themselves. But if time is linear, matters are not so simple. D. H. Mellor, Emeritus Professor of Philosophy at the University of Cambridge, introduces us to the possibilities and impossibilities of time travel under these circumstances. Forward time travel can be accomplished by anyone who can make his clock tick at a slower rate than the clocks in the outside world, a feat which, although perhaps seemingly impossible, can in fact be achieved in either of two ways: by travelling at a speed close to that of light, or, less costly, by slowing down one's metabolic processes. However, backward time travel, requiring as it does the reversal of the time order of events, is fraught with conceptual difficulties. For a time traveller to travel in the true sense of the word to a moment in his past, he must be able to interact with his environment in exactly the same way as any other person present at that moment. However, any time traveller who met this requirement would be capable of causing contradictions. Since this is impossible, the possibility of backward time travel is ruled out.

There is one sense in which we all travel in time, and that is the sense in which time passes ineluctably. Biological organisms measure this passage of time on many different time scales, depending on the length of biologically significant activity cycles. These range from the cycles of spontaneous firings of neurons, lasting no more than a few milliseconds, to the swarming cycles of certain insect pests, which last for several years. One of the most important biological time scales is derived from the so-called circadian cycle, the 24 hour cycle of physiological and behavioural changes shared by almost all higher organisms. What is the genetic basis of this circadian cycle? The remarkable discoveries in this field are described by Charalambos Kyriacou, Professor of Behavioural Genetics at Leicester University. Extensive research over the past three decades has brought to light the striking similarities between the genetic components of the biological clocks of organisms ranging from bread mould and fruit flies to mammals. The current rapid progress in the identification of further clock components is of great importance for the theoretical understanding of the workings of biological clocks. The new insights also open up the prospect of practical applications, such as a method to reset the biological clock of humans. This would greatly alleviate the problems caused by desynchronisation phenomena such as 'jet-lag', as well as the problems experienced by shift workers, whose work rhythms interfere with their biorhythms.

During those hours of the circadian cycle in which we are awake, we carry out a large number of voluntary movements. Even the simplest of these movements involve a complex sequence of events. For example, when we hold up a cup to have it filled with tea, we need to anticipate the effect of the extra weight of the tea in the cup and appropriate action needs to be taken to counteract this effect at exactly the right moment; if we fail to do so we will drop the cup. How does our nervous system organise the timing of such movements? Alan Wing, Professor of Human Movement at the University of Birmingham, reports on the notable advances of recent years in explaining our ability to time our movements. Interestingly, models that had initially been developed to account for the timing of everyday actions have been successfully extended to the timing of much more sophisticated processes such as the generation of musical rhythms. New brain imaging techniques have made it possible to determine which brain regions contribute to the timing of movements, providing valuable knowledge of the consequences of brain damage and neurodegenerative diseases such as Parkinson's disease.

The genetic components of our biological clocks are the same for people the world over, as are the neurological processes underlying the timing of our movements. But the same is not true of the ways in which we talk about time. David Crystal, Honorary Professor of Linguistics at the University of Wales in Bangor, illustrates the dramatic impact of cultural differences on the ways in which time relations are expressed in the different languages of the world. To give but one example: in Western cultures, time is thought of in terms of a line along which we progress, and the multitude of possible relations between points on the time-line is reflected in the grammatical structure of Western languages, for example through the use of different tenses. But not all peoples think of time in terms that can be related to the single dimension of a line. The languages of some of these peoples do make use of tenses, but they are not the familiar tenses of past, present and future. Instead, they might be used to distinguish general truths from reports of known or probable events on the one hand and uncertain events on the other, as for example in the Amerindian language Hopi.

To the Westerner, whose present moment marches inexorably along the time-line and who is given no opportunity to travel into the past, there is but one escape and that is through fiction. Fiction creates a temporal paradox: it suggests plural futures to the listener or reader, who none the less knows that only one of them will be pursued. Gillian Beer, King Edward VII Professor of English Literature in the University of Cambridge, explores the differences between two forms of fiction: storytelling and the novel. A story told aloud is embedded in the real time of its telling; what the story tells may be drawn from the remotest past, but the power of the storyteller lies in the occasion of the telling of the story. For the listener, once the teller has started the story, narrative time unfolds in the same inevitable manner as real time. The relationship between writer and reader is fundamentally different from that between storyteller and listener. Writer and reader are separated in time, and the reader draws into the story circumstances that the writer, locked in the past of writing, cannot control. The reader can read backwards, even sideways. Thus the reader, unlike the listener, inhabits multiple time at will.

Temporality is part of the very essence of our existence. We may think of time as being linear or cyclic or something altogether different, we may contemplate the possibilities of time travel afforded by these different concepts of time, we may speed up or slow down our own internal clocks, but what we cannot do is break free from time. We can live outside time no more than we can live outside the dimensions of space. What of God? Does God exist in time, or is God timeless? This question is taken up by J. R. Lucas, former Fellow and Tutor of Merton College, Oxford. Plato argued that God must be changeless, and, since time implies the possibility of change, he concluded that God had to exist outside of time. Numerous other arguments for the timelessness of God have been developed. Thus it has been said that the creator of all things, including time, cannot Himself exist in time. Similarly, it has been said that the concept of free will can be reconciled with the notion of an omniscient God only if God exists outside time. But when these arguments are examined in detail, they are seen to be based on a misconstrual of the logic of change, and on a confusion between instants and intervals. Once these mistakes are recognised, we no longer feel impelled to think that God must be timeless. However, even if the difference between our existence and God's existence is not that of temporality versus timelessness, it is still true that God's time differs fundamentally from our time: our present interval is of supreme brevity, but God's time is not subject to the limitations of our mortal existence and so God's present interval embraces the whole of time.

1 Time and Modern Physics

CHRISTOPHER J. ISHAM AND KONSTANTINA N. SAVVIDOU

> *Time is a child playing, gambling;*
> *for the kingdom is for the simple.*
> Heraclitus

The subject of time

The subject of 'time' exercises a universal fascination. In no small part this is due to the genuinely interdisciplinary nature of the issues that arise. Thus questions about the nature of time occur in areas as disparate as physics, biology, psychology, philosophy, poetry (think of the work of T. S. Elliot), visual art, theology, music (for example, in the chanting of plainsong) and many more.

Some of these topics are covered in other chapters in this book, but in all cases – or, at least, in the more academic disciplines – a basic question is how the concept of time fits into the underlying metaphysical structure of the subject concerned. Thus, for us, a key issue is the role played by time in the foundations of modern physics. And, as theoretical physicists, we are particularly concerned with how the answer to this question relates to the various *mathematical* structures that are involved in the physicist's account of time.

Let us begin by remarking that there are two quite different ways in which time has been viewed by physical scientists: these are known as the *absolute* and *relational* ideas of time. In essence, the difference comes down to whether or not we grant time (and space) an existence independent of material objects and processes. According to the absolute view of time (and space), time (and space) simply form the 'arena' of physics: the background structure within whose framework all of physics is necessarily phrased. On this view, material processes take place against the background of an independent 'something' called time (and space). Newtonian physics and the theory of special relativity are good examples of theoretical frameworks of this type.

On the other hand, the relational view denies time (and space) an existence independent of material objects and processes. On this view, time exists only by virtue of the existence of matter and material events. Thus the concept of time is dependent in some way on the idea of matter. This view is famously associated with the names Leibniz and Mach. General relativity is arguably a theory of this type, although in its typical applications it also assumes certain absolute structures.

To a significant extent, modern physics oscillates uneasily between these two perspectives on time. One important issue is the way in which they are related. For example, what is the role of a 'clock' in this respect? Let us consider a wristwatch. On the one hand, it is made of matter and in that sense its temporal qualities are naturally associated with the second view. On the other hand, when we talk of a 'good' watch, we typically mean the extent to which it *measures* accurately the background absolute time of Newtonian physics – a concept that clearly accords with the first view of space and time.

But what about an atomic clock, which also is made of (quantum) matter? In this case we typically talk about the clock *defining* time, rather than measuring it. But what then is meant by a 'good' atomic clock: are some definitions of time 'better' than others; and how are they related to each other, and to the background time of Newtonian physics?

Clearly, one format for a chapter on 'Time and Modern Physics' would be to survey the different ideas of time in the classical and quantum versions of Newtonian physics, special relativity and general relativity. However, we have elected to follow a different route and to concentrate instead on two particular ways in which time arises in modern physics: as the parameter in temporal logic and as the parameter of dynamics. This will allow us to touch on many of the basic ideas concerning time, as well as to discuss some very recent ideas about time in physics.

Two roles for time

The nature of time is something that much occupied St Augustine: perhaps because, as he explains in his *Confessions*, he had such a 'good time' when he was a young man! Whatever the case may be, it is appropriate in our case to start with the following well-known excerpt from *The Confessions*:

> What then is time? If no one asks me I know; if I want to explain it to a questioner; I do not know. But at any rate this much I dare affirm I know: that if nothing passed there would be no past time; if nothing were

approaching there would be no future time; if nothing were then there would be no present time.

The ideas implied here are as profound and relevant today as when the saint first stated them. One such is the universally acknowledged fact that 'time' is an elusive concept: in one sense we think we know exactly to what it refers, but when we try to pin it down it slips away – like a chimera, a will-o'-the-wisp. However, of more direct relevance to our present task is the several different roles for time that are implicit in Augustine's remarks. This theme is anticipated in the earlier comments of Aristotle, in his *Physics*, on the notion of time:

> But when we perceive a distinct before and after, then we speak of time; for this is just what time is, the calculable measure or dimension of motion with respect to before-and-afterness. Time, then, is not movement, but that by which movement can be numerically estimated. And as motion is a continuous flux, so is time; but at any given moment time is the same everywhere, for the 'now' itself is identical in its essence, but the relations into which it enters differ in different connections, and it is the 'now' that marks off time as before and after. But this 'now' which is identical everywhere, itself retains its identity in one sense, but does not in another; for inasmuch as the point in the flux of time which it marks is changing the 'now' too differs perpetually, but inasmuch as at every moment it is performing its essential function of dividing the past and future, it retains its identity.

In a related context, in the Greek Orthodox Church – which inherited the tradition of Greek philosophy – it is believed that, together with faith, the study of nature ($\phi v \sigma \iota \xi$) leads us to God. At the basis of this study lie what the Greeks called the 'categories', which are the properties without which nothing can be perceived or encompassed. According to Aristotle they are ten: substance, quantity, quality, relation, place, time, space, to have, to act, to be acted upon. Maximus the Confessor (one of the Fathers of the Orthodox Church) taught in regard to the category of time:

> The beginning, the middle, and the end are features of all that can be divided in time, and it is also true to say, of all that we can perceive inside eternity. For time, because it has movement that can be measured, is specified numerically. Eternity, because together with existence it incorporates the category of time, has one dimension because it contains the origin of being ($o v$). If time and eternity are not without a beginning, even more are all that are contained in them.

(Maximus, *Theological and Philosophical Questions*)

It is evident that Maximus discriminates between time as motion that can be measured, and time as eternity, without the notion of change.

Let us now turn to the physical sciences, with the observation that a very important feature of time in physics is that the 'way things are' (more technically, the state of a system, see later) is specified at a given moment of time. In particular, in classical physics, at any given time t any proposition about the system is either true or false. Furthermore, such propositions can be combined using the operations of standard logic. Thus, if A and B are a pair of propositions, we can construct 'A and B', 'A or B', 'A implies B'; and for any single proposition A, there is the negation '*not A*'.

In this context, we can read into the remarks of Augustine, Aristotle and Maximus the following two roles that 'time' plays in physics:

1. Time appears as an ordering parameter in the sense that there is a separation of temporal experience into the 'past', the 'present', and the 'future'. This ordering of 'states of being' gives rise to the concept of *temporal* logic. Thus, for example, if A_{t_1} is a proposition about the state of the system at time t_1, and B_{t_2} is a proposition referring to the system at a later time t_2, we can form the temporal conjunction 'A_{t_1} and then B_{t_2}'. From this perspective, the ordering parameter in temporal logic is time viewed from the perspective of 'being' – a time must be specified in order to say 'how things are'.
2. Another facet of time is to view it from the complementary perspective of 'becoming' – the idea of time arises in saying 'how things change'. Thus time appears as the parameter of evolution that arises in the description of the dynamics of a physical system.

From the viewpoint of theoretical physics, it is important to understand the way in which these two aspects of time are represented mathematically in the relevant equations. In particular, we shall consider carefully the way in which real numbers are used in these two roles.

In the context of this chapter, we shall concentrate mainly on how these two roles manifest themselves in classical physics – especially in regard to the background, absolute time of Newtonian physics. Of course, in accordance with the developments of this century, one should really consider time in the context of special – and perhaps even general – relativity, in which case the relevant subjects are 'causal logic', and 'relativistic dynamics'. For our purposes it will not be necessary to make this extension in detail. In later sections, however, we will say something about the famous 'problem of time' in the quantum theory of general relativity as this throws into doubt the fundamental status of all our standard ideas about time.

Time ordering and real numbers

The temporal ordering of events as 'past', 'present' and 'future' is not the only way in which temporal concepts have been construed in different cultures and ages. Thus, for example, the idea of 'circular' time arises in a variety of myths of eternal recurrence, in particular in the thought of ancient Greece, and of India (see Thapar, Chapter 2, this volume). If time is represented mathematically by a circle then it is clear that no real concept of history can be developed. For if an event lies in the future of a present one, then it also lies in its past.

This circular view of time contrasts sharply with the modern, linear picture, which it is often argued, has its roots in Judaeo-Christian theology. This world-view is fundamentally historical, the key events being strung out between the creation of the world and its final apocalyptic consummation. In Christianity, this linear ordering of events is additionally *centred* on the birth of Christ: every occurrence is either 'before Christ' (BC) or 'in the year of the Lord' (anno domini, AD).

Thus, of any pair of events, labelled E and F, Judaeo-Christian sensibility asserts that F lies in the future of E, or E lies in the future of F, or E and F are contemporaneous. This explains why the real numbers (which can be construed as distances along a straight line, measured from an arbitrarily fixed zero point) are a natural mathematical model for time, since for any pair of real numbers a and b it is true that $a<b$, or $b<a$, or $a=b$.

To understand more fully the role of the real numbers in the representation of time – both in regard to temporal logic and to dynamics – it is necessary to look more closely at the type of mathematical structures that are used in theoretical physics, in particular the idea of the space of states of a system. We will concentrate mainly on the situation in classical physics, and defer discussion of quantum theory until later sections.

Classical physics: the logic of propositions and the space of states

Consider an arbitrary physical object, or physical system, at some moment of time. In classical physics, a key feature of such a system is that it is completely defined by specifying all its properties at that moment – what we have referred to above as the time of 'being'. Such a defining list of properties is called a *state* of the system. Of course, although at any given moment of time a physical system is in one, and only one, physical state (this is implicit in the concept of 'state'), at different moments of time the system can (and generally, will) be in different states. The set of all possible states of a system is called the *space* of states, or *state space*.

To illustrate these ideas, consider the following example. Imagine a point particle (a theoretical entity of no spatial extension but with a fixed mass) moving in one dimension (i.e. along a line) under the influence of forces of a Newtonian type. It can be shown that such a system is completely defined by two properties: the position (denoted by x) and the momentum (mass × velocity; denoted by p) of the system. In other words, the state of such a system is completely determined by specifying the values of the position x and momentum p of the particle; thus the space of states of the system is a two-dimensional space with coordinates x and p, as shown in Figure 1.

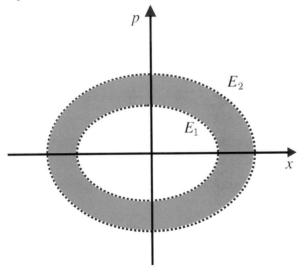

FIGURE 1. A classical statespace.

Of course, the point particle will have other physical properties besides position and momentum; for example, it will have a certain energy (denoted by E). However, the energy of the point particle is determined by its state, i.e. by the values of the position and momentum: mathematically – in the particular example in Figure 1 – we have

$$E(x,p) = \frac{p^2}{2m} + kx^2, \qquad (1)$$

where m denotes the mass of the particle, and k is some positive constant. From this definition of the energy of the point particle, it follows immediately that different states can give rise to the same value of the energy. This prompts the following question: what is the set of all states (x,p) for which the proposition 'the energy of the system has the value E_1' is true? This set of states is represented by the inner ellipse in Figure 1. Similarly, the outer ellipse represents the set of states for which the proposition 'the energy of the system has the value E_2 (with $E_1 < E_2$)' is true. The proposition 'the energy of the system lies between E_1 and E_2' is represented by the shaded subset between the two ellipses. In fact, *every* proposition about the point particle can be represented by a set of states, namely the set of states for which the proposition is true.

This idea can be generalised to *all* classical systems: if we denote the state space of such a system by S, then *every* proposition P about the system can be represented by an associated subset S_P of S – namely the set of all states for which the proposition is true. Conversely, every subset of S represents a proposition: more precisely, every such subset represents *many* propositions, each of which asserts that the value of a certain physical quantity lies in a certain range.

It is easy to see how the logical calculus of propositions is represented in this picture. For suppose that P and Q are a pair of propositions, represented by the subsets S_P and S_Q, respectively, and consider the proposition 'P and Q'. This is true if, and only if, both P and Q are true; and hence the subset of states representing this logical conjunction are those states that lie in both S_P and S_Q, i.e. the set-theoretic intersection $S_P \cap S_Q$. Thus 'P and Q' is represented by the subset $S_P \cap S_Q$ of the state space S.

Similarly, the proposition 'P or Q' is true if either P or Q (or both) are true; and hence this logical disjunction is represented by those states that lie in S_P *plus* those states that lie in S_Q, i.e. the set-theoretic union $S_P \cup S_Q$ of the two sets. Finally, the logical negation 'not P' is represented by all those points in S that do *not* lie in S_P, i.e. the set-theoretic complement S/S_P.

In this way, a fundamental equivalence is established between the logical calculus of the propositions about a physical system, and the Boolean algebra of subsets of the associated space of states.

Dynamics and the representation of time by real numbers

We can now use the idea of the space of states S to discuss the way in which dynamical evolution is handled in physics. The key idea here – at least, in classical Newtonian physics – is that the state of a system changes in a deterministic way with respect to the background Newtonian time t. What this means is that if s_{t_1} is the state of the system at some time t_1, then the state s_{t_2} at any later (or earlier) time t_2 is uniquely determined by the forces acting on the system.

The precise way in which s_{t_1} changes into other states as time evolves is given by Newton's famous second law of motion:

$$F = m \times a, \qquad (2)$$

where F is the force acting on the system, m is the system's mass and a is the acceleration. For our point particle moving in one dimension, this equation is equivalent to the following set of so-called (first-order, ordinary) *differential equations*:

$$m\frac{dx}{dt} = p \qquad (3)$$

$$\frac{dp}{dt} = F. \qquad (4)$$

Thus we see how the real numbers arise again as a model for time: namely as the independent parameter in the theory of ordinary differential equations. Hence, ultimately, this use of the real numbers depends on their role in *differential calculus*. This role is related, but not equivalent, to their appearance as the ordering parameter in temporal logic. Indeed, one of the main points we wish to make is that for the future development of theoretical physics in, say, quantum gravity, it may be profitable to distinguish between these two roles more sharply than is usually the case. We shall return to these issues in later sections.

Temporal logic in this approach

Let us consider now how the temporal-logic proposition 'P at time t_1 and then Q at time t_2' might be represented in this formalism, where (say) $t_1 < t_2$.

In our previous discussions, we saw that the proposition 'P and Q' could be represented by the subset $S_P \cap S_Q$, which comprises all those states for which P is true *and* Q is true. But notice that in the case of the proposition 'P at time t_1 *and then* Q at time t_2' we must proceed more carefully, since we are comparing states at different moments of time.

We can suppose that, as in the discussion above, the propositions about the system are represented by subsets of the state space S. Furthermore, if we think of the states in S as being the states of the system at time t_1 (thus we chose t_1 as an initial 'reference' time), then the proposition 'P at time t_1' corresponds to the subset S_P. The main step now is to find the mathematical representation of the proposition 'Q at time t_2' as a subset of the set S of states at time t_1.

In this context, we recall from the discussion above that states evolve deterministically, which means that every state at t_1 has a unique successor state at t_2; conversely, every state at t_2 has a unique predecessor state at t_1. In particular, each state s at time t_2 that makes Q true (i.e. s lies in S_Q) has its own unique predecessor state at t_1. We will denote by $(S_Q)_{\text{pred}}$ the subset of S of all such predecessors of states that lie in S_Q: clearly, this is the subset of S that represents the proposition 'Q at time t_2'.

Of these predecessor states, some will lie in S_P, and others will not. Thus the subset that represents our proposition 'P at time t_1 *and then* Q at time t_2' consists of those predecessor states in $(S_Q)_{\text{pred}}$ that also lie in S_P: namely, the set-theoretic intersection

$$S_P \cap (S_Q)_{\text{pred}}. \tag{5}$$

This is a perfectly workable definition and it is implicit in the usual treatment of classical mechanics, but it has the feature that the set-theoretic representation of the temporal-logic proposition depends explicitly on the detailed *dynamics* of the system (via the evolution of subsets of S into other subsets). From our perspective this is undesirable, since it mixes completely the two uses of time: what we have called the time of being and the time of becoming. And, anyway, it is anomalous in the sense that the physical *meaning* of the proposition 'P at time t_1 *and then* Q at time t_2' is independent of the dynamics, and hence it is natural to require that there be a mathematical representation of the proposition with the same property. This can indeed be done, as we shall now see.

A history version of classical physics

It may seem surprising, but it seems to have been only very recently that the possibility of finding a dynamics-independent representation of temporal logic was discussed. In fact, the question arose first in the context of quantum theory; only after the issue was discussed there was it realised that there is an analogous construction in the classical case (as discussed by K.N.S. in 1999).

The mathematical framework involved in the case of classical physics is illustrated in Figure 2. The key idea is to start *ab initio* with two time variables – a time of 'being' and a time of 'becoming' – and then to associate with each time of being t a *separate* copy S_t of the classical state space, which is to be thought of as representing the states of the system at that time t.

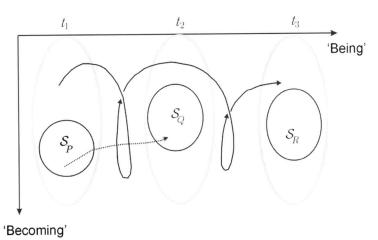

FIGURE 2. History formalism for classical physics.

A natural mathematical quantity associated with this construction is a *path*: this a function γ that, for each time t, assigns a state $\gamma(t)$ that lies in the state space S_t. From a physical perspective, each such path γ describes a possible *history* of the system: i.e. $\gamma(t)$ describes 'how things are' at the 'time of being' t.

A proposition like '*P at time t_1 and then Q at time t_2*' is then represented by the set of all *paths* γ with the property that at the time t_1, the state $\gamma(t_1)$ lies in the subset S_P of S_{t_1}, and at the time t_2, the state $\gamma(t_2)$ lies in the subset S_Q of S_{t_2}. This representation of temporal propositions by subsets of paths is clearly independent of dynamics, and it is straightforward to check that it reproduces the expected logic of temporal propositions.

The idea of dynamics arises in an interesting way in this framework. From a mathematical perspective, there is a dynamical evolution associated with *each* copy S_t of the state space: thus the time of temporal logic and the time of dynamics are now separate from each other.

This is reflected in the fact that there are *two* types of time transformation in this formalism: one that transforms the 'external' time indices on temporal propositions (the time of 'being') and one that is associated with the internal evolution (the time of 'becoming') in each S_t for every value t of the external time label t. However, it is must be emphasised that we are *not* claiming that 'physical' time should be represented by a two-dimensional space (in addition to the three dimensions of physical space; so that space-time would be five-dimensional): in fact, for any given physical system the two types of time transformation are locked together, and the actual history of the system is a single path, as represented by the spiral in Figure 2.

It came as a great surprise to us to discover that the central idea behind the two types of time transformation had to some extent been anticipated by the artist John Latham. In a series of works known generically as 'time-base rollers' he has presented a graphic representation of these ideas: an example is the picture in Figure 3. The axis going horizontally along the length of the roller (what Latham refers to as the 'time-base') is the analogue of our 'time of being', while the dynamical evolution associated with each such time is represented by the vertical motion of the blind as it is unrolled.

From a philosophical perspective, certain ideas of process philosophy underpin Latham's general metaphysical views, and it is interesting to ask

FIGURE 3. One of John Latham's time-base roller constructions.

how process ideas might be reflected in the theoretical physicist's representation of time.

The mathematics of process

We shall now address once more the important question of how the 'time of being' and the 'time of becoming' are related to each other in theoretical physics. In particular, we want to consider how one might mathematically implement the vision of a process philosopher such as A. N. Whitehead in which the notion of the time of 'becoming' is structurally independent of the time of being, whereas the converse is normally the case (see below). Indeed, Whitehead emphasises that, from a purely empirical perspective, the notion of 'event' – and the relation between events – is more fundamental, and our normal concepts of space and time should be understood as idealised mathematical constructs from the empirical data pertaining to events.

This is reflected in his concept of the *discernible*, which is to be viewed as everything that happens in some (time) *duration*. The concept of a 'duration' is irreducible in the sense that it is *not* to be construed as a collection of 'moments of time'. A key property of durations is their ability to contain one another, and it is this type of ordering property that Whitehead considers to be fundamental. On the other hand, there is the notion of *instantaneousness*, which is an idealised, logical concept of all nature at an instant, where an instant is conceived of as deprived of all temporal extension; from this concept also arises the idea of a 'moment' of time.

Thus we are interested in finding a mathematical representation of 'durations' that does not depend on the idea of a 'moment' of time (which is represented normally by a real number). To this end, let us consider again the way in which the time of becoming is represented mathematically in standard classical physics. This involves considering the mathematical structure of the dynamical laws of classical physics.

The dynamical equation of elementary classical physics is Newton's second law of motion, which, as we have seen before, can be written as the coupled differential equations

$$m\frac{dx}{dt} = p \qquad (6)$$

$$\frac{dp}{dt} = F. \qquad (7)$$

For our purposes it is important to note that the time derivative dp/dt in equation (7) is to be interpreted mathematically as the limit

$$\text{Limit}\, \frac{p(t') - p(t)}{t' - t} \qquad (8)$$

as t' gets closer and closer to t. A similar remark applies to the time derivative dx/dt in equation (6).

Now, as we emphasised earlier, the 'time of becoming' variable arises mathematically via the theory of differential equations and differential calculus, as in equations (6) and (7). On the other hand, the quantities t' and t in equation (8) are points of the 'time of being', since they label the specific times at which the momentum variable p has a value. This is the sense in which the concept of 'becoming' is structurally dependent on the concept of 'being'.

From a mathematical perspective, the reason for this dependence is that there are no genuine infinitely small numbers (so-called 'infinitesimals') in standard mathematics, and hence the symbol 'dt' has to be interpreted in a limiting sense, as in equation (8). However, there is a subject known as synthetic differential geometry in which genuine infinitesimals *do* exist. This suggests that dynamics could be formulated in a new way in which the concepts of 'becoming' and 'change' are independent of any underlying points of temporal being; specifically, one could try to identify the infinitesimals as the mathematical analogues of Whitehead's 'durations'.

By these means one would gain an implementation in mathematical physics of certain ideas of process philosophy. However, in doing so it is necessary to introduce a non-standard model of the real numbers, and – it transpires – an underlying logical structure that is *intuitionistic*. In standard logic we have the principle of excluded middle, which asserts that the disjunction of any statement with its negation is always true. In intuitionistic logic, contrary to standard logic, the principle of excluded middle no longer holds, i.e. propositions of the form '*P* or (not *P*)' are not necessarily valid. We note *en passant* that the inapplicability of the principle of excluded middle is a characteristic feature of the so-called *constructive* approach to mathematics in general.

Time in quantum theory

There are analogues in quantum theory of the various aspects of time in classical physics that we have discussed above, albeit complicated by the special role of the idea of 'measurement' in the standard interpretation of the theory.

Quantum theory is of fundamental importance in describing the world at atomic and subatomic scales. However, it differs from classical physics in one very important respect: in classical physics, in any specific state, a proposition such as 'the particle is at position x' is either true or false. On the other hand, in quantum physics, the best that can be done is to talk about 'the probability that, *if* a measurement of position is made, the particle will be found at x', and this probability is assigned a numerical value between 0 and 1.

This strict instrumentalism that lies at the heart of the standard interpretation of quantum theory, means that the concept of 'being' is replaced by that of 'being measured'; which leads one to wonder what happens to the distinction made earlier between the time of temporal logic ('being') and the time of dynamics ('becoming')?

In regard to dynamics, the answer is that – provided no measurements are made – quantum states, and their associated probabilities, evolve deterministically with respect to a background Newtonian time: so in this respect classical and quantum theory are alike. In particular, the dynamical evolution of the mathematical object (a vector in a vector space) that represents a quantum state is described by a differential equation (the Schrödinger equation), and the remarks made earlier about the implications of this for the mathematical model of time apply here too.

However, in the standard interpretation of quantum theory there is another type of time evolution that can occur. This arises if a measurement of some physical quantity is made: the quantum state is then deemed to change instantly into a different one that reflects the *actual* result obtained from among the set of those that are probabilistically possible according to the theory. Many physicists feel this to be one of the most unsatisfactory features of standard quantum theory, and there have been many approaches to the task of developing a theory in which these sudden apparent 'collapses' of the state vector can be derived in some way; for example, from the deterministic evolution of the state vector of the system that includes the measuring device, now regarded as a quantum entity in its own right. One such scheme – the *consistent histories* approach to quantum theory – bears directly on our topic of the two facets of time.

In regard to temporal logic, the standard quantum formalism certainly allows the assignment of a probability to getting a particular sequence of results of measurements of various observables made at a sequence of times, but this depends explicitly on the details of the dynamics of the system, and thus we once again have a situation in which the time of temporal logic is mixed with the time of dynamics.

What is needed to separate these two aspects of time is a quantum analogue of the history formalism discussed above for the classical case. But to do that it would be necessary to work with a version of quantum theory that deals with sequences of *values* of quantities, not results of the *measurements* of quantities – something that is not possible in standard quantum theory. However, in recent years a new approach to quantum theory has been developed, known as the *consistent histories theory*, which is aimed at this precise point. The penalty extracted by the quantum realm for being allowed to talk about sequences of values (rather than results of measurements) is that probabilities can be assigned only to a limited class of propositions.

This theory has been developed by ourselves and co-workers in such a way as to place emphasis on the idea of quantum temporal logic. In this form it is relatively easy to implement the idea of two times and the associated two types of time transformation: indeed, as remarked earlier, it was in the context of this theory that the idea of the two types of time variable first arose – only afterwards was it appreciated that there is an analogous structure in classical physics.

It is tempting perhaps to connect the two types of time evolution in standard quantum theory (deterministic dynamics, and the collapse of a state vector when a measurement is made) with the two types of time transformation that arise in the history theory. Indeed, one of us (K.N.S.) has conjectured the existence of a close correspondence between the collapse of the state vector and changes in the time of 'being'.

Spacetime, gravity and a variety of times

The central idea of special relativity is that the three-dimensional space and one-dimensional time of Newtonian physics are combined together to give a single, four-dimensional structure known as *spacetime*. This is a radical step but, nevertheless, the role thereafter of spacetime is not dissimilar to that played in Newtonian physics by the separate concepts of space and time. In particular, the geometry of the spacetime structure (which determines things like the distance between two points in spacetime) is fixed and forms the background within which the dynamical equations of physics are framed. (It is not just fixed, but also relatively simple: we talk of the spacetime in special relativity being 'flat'.)

The situation in general relativity is quite different. The geometry of spacetime is no longer fixed, but instead depends on the energy and matter present in the universe. This dependence is captured by Einstein's famous field equations, which describe the precise manner in which the 'geometric field' (which corresponds physically to the gravitational field) depends on the distribution of matter and energy in the universe. The result of this dependence is that spacetime is no longer 'flat', but instead becomes 'curved'.

The idea that spacetime is curved has a radical effect on our concept of time. Einstein's field equations have many different solutions, leading to many different spacetimes. Many of these have the special property that the four-dimensional spacetime can be regarded as a 'stack' of curved three-dimensional spaces, each of which is given a unique label that can be thought of as the 'moment' of time corresponding to that three-dimensional space. In this sense, a four-dimensional spacetime can be thought of as a history of three-dimensional space. The events occurring in the four-dimensional spacetime can be time-ordered using the labels of the three-dimensional spaces in which they are located. In this respect, the picture looks similar to that of Newtonian physics.

However, a key feature of general relativity is that the manner in which a four-dimensional spacetime is divided into a stack of three-dimensional spaces is very non-unique. In fact, almost any way of dividing the spacetime can be chosen provided only that the ensuing stack of three-dimensional spaces has the property that no light beam can travel across any single one of them. Since nothing travels faster than light, the key idea here is that any entity must move from one element of the stack to the other as 'time' increases.

All these different ways of carving up four-dimensional spacetime into three-dimensional spaces result in different time orderings of events. Whether or not any particular carving-up is admissible, and yields a well-defined time, depends on the geometry of spacetime and hence on the distribution of matter in the universe. This picture is far removed from that of the fixed, universal time of Newtonian physics.

The problem of time in quantum gravity

The central feature of time in quantum gravity arises from the fact that, as emphasised above, in general relativity what constitutes an admissible way of labelling events as occurring at different times depends on the geometry of spacetime. However, in a theory of quantum gravity (which combines the theory of general relativity with quantum theory) we expect this geometry to have the same type of quantum features as we discussed above: in particular, the geometry will not have a definite value, but will be determined only probabilistically. But if the spacetime geometry had no definite value, then presumably neither would the set of allowed ways of introducing time:

in particular, there would be no single choice that can serve as a uniform way of labelling events as occurring at different times for all the spacetime geometries that could appear with non-zero probability.

A detailed study of what is known as the 'canonical' quantisation programme confirms these expectations: there is indeed something like a probabilistic distribution of three-dimensional geometries, but with no time label at all! Making sense of this peculiar situation is known as the *problem of time* in quantum gravity.

Many different ways have been suggested for recovering the notion of time in quantum gravity. One possibility is that there is some universal choice for time that stands outside the internal structure of general relativity. This suggestion tends to be particularly popular with process philosophers and theologians who have an interest in theoretical physics. However, since it seems to violate the spirit – if not the law – of general relativity it does not appeal to the majority of those who work in quantum gravity.

In practice, most research in canonical quantum gravity has appealed to the notion of 'internal' time in which part of the gravitational field is used as a local clock to specify the time at which the remaining parts of the gravitational field have a probabilistic distribution. This relational view of time has attractive features but it also raises some difficult questions. In particular: (i) can a choice for internal time be found such that the ensuing dynamical equations for the rest of the field are of the type encountered elsewhere in quantum theory; and (ii) if this is the case, can the predictions associated with different such choices (assuming there are such) be related in a physically meaningful way?

Unfortunately, the prognosis for answering 'Yes' (or, at least, nearly 'Yes') to these questions is not good. The evidence suggests that the dynamical equations agree only approximately with standard ones: a situation that has led to suggestions that what we normally call 'time' will emerge from the formalism only in a coarse-grained way (analogous, perhaps, to the ideas of temperature or pressure in the theory of a gas). This throws into severe doubt the validity of the probabilistic structure as a fundamental ingredient of the theory, and hence the viability of the quantum formalism as a whole when applied in the context of quantum gravity. Also, model calculations suggest that different choices of internal time would lead to different families of probability distributions, and in the absence of any background spacetime reference system it is difficult to see how the different results could be compared physically.

Under these circumstances, we may well wonder what happens to our two different concepts of time: the time of temporal logic ('being'), and the time of dynamics ('becoming')! Our suspicion is that, in order to answer such questions, it will be necessary to introduce radical revisions in the basic ideas of quantum theory itself.

Zervan and the creation of time

One can often find silver linings in life, and in the present context the obscurities in the concept of time in quantum gravity allow the possibility of developing theories of *cosmogenesis* that deal with the origin of the universe itself, including the quixotic concept of the 'origin of time'. The most famous quantum cosmology theory of this type is that due to Jim Hartle and Stephen Hawking.

The problem of the 'origin of time' has preoccupied certain earlier civilisations. For example, the ancient Persian religion of Zoroaster conceived of a primordial god, Zervan (see Figure 4), who existed by, and in, himself – εv $\alpha \rho \chi \eta$ – before the origin of time. Feeling alone, he wished to conceive in order to have a companion, and being alone, but also being a God, he was obliged to do this by offering sacrifices to himself that he might have a son. But, during this process, he began to doubt if his sacrifices would work, and from this doubt there eventually came to be another son, Ahriman – the principle of negation – who was born before the good son, Ohrmazd, whom Zervan desired.

Being fair minded, Zervan had to allow his first-born son Ahriman to rule for a while, but he also decreed that he should be overcome by Ohrmazd, the power of good. And to allow this battle to take place, Zervan created the physical world, and what we call 'time' as the background against which this cosmic struggle can occur. Note that a common gnostic theme is that evil – which is also often associated in some way with the material world – arose as a hypostasis of the doubt of a high spiritual being.

It is interesting to note the ascending spiral in the picture of Zervan. Bearing in mind our earlier remarks about spirals and dynamics, it is clear that the ancient Persians also knew about the two times!

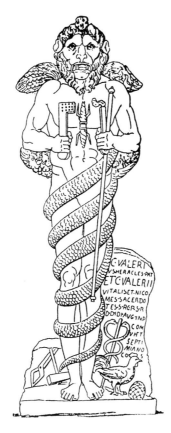

FIGURE 4. The god Zervan.

Conclusions

We have argued that there are two main uses for the concept of time in physics: (i) as the parameter in temporal logic, labelling the points of 'being'; and (ii) as the parameter in the equations of dynamics, where it refers to the notion of 'becoming'.

The full implementation of the ideas of temporal logic in a dynamics-independent way requires a *history* formulation of physics, and we have illustrated briefly how this arises in the classical case.

We have emphasised that a key question for theoretical physics concerns the appropriate mathematical structure that is used to represent the idea of time. In particular, the real numbers appear as: (i) the ordering parameter that labels the points of being; and (ii) the parameter that pertains to becoming, via dynamics with its use of differential calculus and the theory of differential equations. The standard way of defining differentiation – via a limiting procedure – necessarily mixes up the two ideas of time. However, there are other models of real numbers – in particular, in synthetic differential geometry – which allow for genuine infinitesimals, and which hence offer a way of separating mathematically the two concepts of time.

Then we discussed briefly the deep question of time in quantum gravity, and the possibility that what we normally call 'time' is not a fundamental concept but rather one that arises only in some sort of coarse-grained way. Ideas of this sort give rise to the possibility of constructing quantum theories of the origin of the universe and, in particular, the 'beginning' of time.

Finally we note that in the course of this chapter we have introduced ideas from physics, mathematics, philosophy, theology, visual art and Persian gnosticism. If nothing else, this should suffice to demonstrate the remark made in the beginning of this chapter about the interdisciplinary nature of interest in time!

FURTHER READING

Corbin, H., 'Cyclical time in Mazdaism and Ismailism', in *Man and Time: Papers from the Eranos Yearbooks 3*, Princeton, NJ: Princeton University Press, 1973.

Eliade, M., *The Myth of the Eternal Return*, Princeton, NJ: Princeton University Press, 1974.

Gish, N. K., *Time in the Poetry of T.S. Eliot*, London: Macmillan, 1981.

Isham, C. J., 'Creation of the Universe as a quantum tunnelling process', in *Our Knowledge of God and Nature: Physics, Philosophy and Theology*, ed. R. J. Russell, W. Stoeger and G. V. Coyne, pp. 374–408, Notre Dame: University of Notre Dame Press, 1988.

Isham, C. J. and Butterfield, J. B., 'Spacetime and the philosophical challenge of quantum gravity', in *Physics Meets Philosophy at the Planck Scale*, ed. C. Callender and N. Huggett, pp. 33–89, Cambridge: Cambridge University Press, 2000.

Savvidou, K. N., 'The action operator in continuous time histories', *Journal of Mathematical Physics* **40** (1999), 5657–5674.

Whitehead, A. N., *Concept of Nature*, Cambridge: Cambridge University Press, 1978.

Chapter 2, Plate I Painting on cloth from western India, representing the idea of genealogy (eighteenth century).

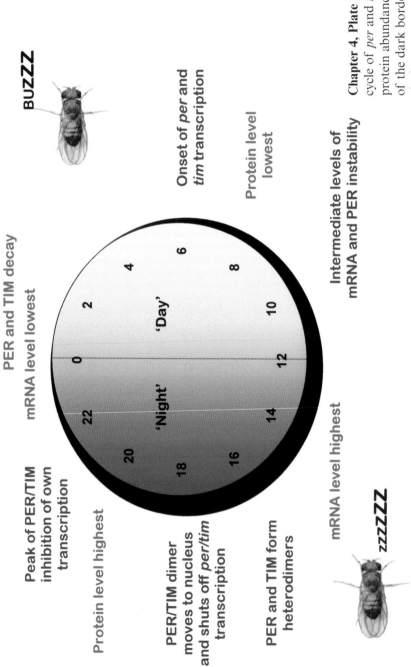

Chapter 4, Plate I The molecular cycle of *per* and *tim* mRNA and protein abundance. The thickness of the dark border around the clock represents the levels of *per* and *tim* mRNA which are at their peak early at night. The major molecular developments are listed at approximately the position in the circadian cycle at which they occur.

Chapter 4, Plate II The PER-TIM negative feedback loop in the pacemaker cells of *Drosophila*. The various molecular players in the circadian drama are illustrated. The genes are shown as black lines with arrows; ccg refers to clock-controlled genes. The phosphorylation of PER by the kinase DBT is represented by the green circle carrying the letter P. The red arrows show the inhibition of the positive transcription factors (+) CLOCK-CYC and the block on *per/tim* transcription by the negative factors (−) PER-TIM. Inset is shown the activation of CRY by light and the light-driven degradation of TIM during the nuclear phase. This results in a derepression of *per* and *tim* transcription when the negative effects of the PER-TIM dimer on CLOCK-CYC are lifted.

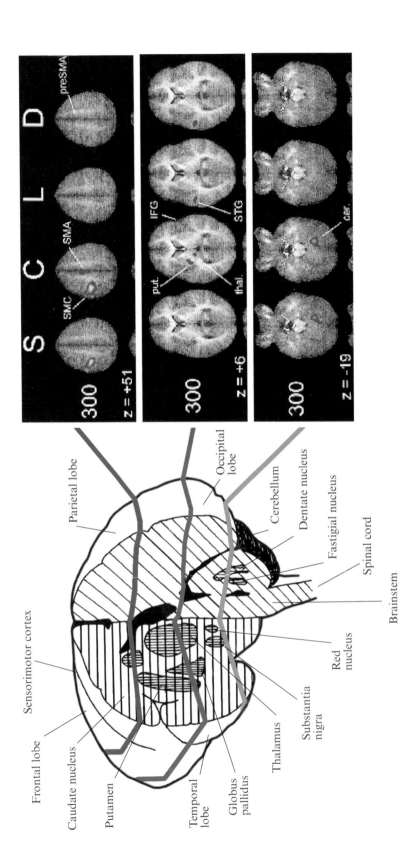

Chapter 5, Plate I Brain activation in synchronisation and continuation timing. The 'slices' moving progressively down the brain (decreasing z height) show elevated activity in areas marked orange/red that differ according to condition S (synchronisation), C (continuation), L (passive listening control), D (pitch discrimination control). SMC, sensorimotor cortex; SMA, supplementary motor area; preSMA, cortex anterior to the SMA; put., putamen; thal., thalamus; IFG, inferior frontal gyrus; STG, superior temporal gyrus; cer., cerebellum.

2 Cyclic and Linear Time in Early India

ROMILA THAPAR

Introduction

The received wisdom of the past 200 years describes the traditional Indian concept of time as cyclic, excluding all other forms and incorporating an endless repetition of cycles. This was in contrast to what was perceived as the essentially linear time of European civilisation. Implicit in this statement is also an insistence that cyclic time precludes a sense of history, a view which contributed to the theory that Indian civilisation was ahistorical. Historical consciousness it was said, required time to be linear, and to move like an arrow linking the beginning to a final eschatological end. Concepts of time and a sense of history were thus interwoven.

Early European scholars working on India searched for histories of India from Sanskrit sources but were unable to discover what they recognised as histories. The exception was said to be Kalhana's *Rajatarangini*, a history of Kashmir written in the 12th century. It is indeed a most impressive pre-modern history of a region, but it is not an isolated example, since this genre finds expression in other regional chronicles, even if the others were not so impressive. These others were ignored, perhaps because they were less known to European scholarship; perhaps because if Indian civilisation were to be characterised by an absence of history it would become all the more necessary that Indian history be 'discovered' through the research of scholars who came to be labelled as Orientalists.

Some among these scholars did suggest that there was a strand of linear time in certain texts, but the dominant view remained that of insisting on the dichotomy of cyclic and linear, and reiterating that the Indian time concept was cyclic. Another aspect of this was the derision over the length of the cycle, which ran into well over a million years. For those observing the

18th century Irish cleric Bishop Usher's calculation of about 6000 years for the life of the universe, the figures given in Indian sources seemed absurd. But this derision changed to puzzlement when geology indicated that the universe had existed for a few million years.

Cyclic time in India, endlessly repeating itself and with no strongly demarcated points of beginning or end, was said to prevent a differentiation between myth and history, and to deny the possibility of unique events that are a precondition to a historical view. Repeating cycles would repeat events. This minimised the significance of human activities. The construction of the cycle was said to be a fantasy of figures intended to underline the illusory nature of the universe. Nor was there any possibility of suggesting that historical events in India were moving towards the goal of 'progress', an idea of central concern to 19th century Europe. In the discovering of the Indian past, the premise of investigation remained the current intellectual preconceptions of Europe.

These preconceptions projected Asia, and particularly India, not only as different from Europe but essentially a contrast to Europe. Asia was the 'Other' of Europe. If Karl Marx and Max Weber were looking for contrasting paradigms in understanding the structure of the Asian political economy or the function of religion in Asia, lesser thinkers – but influential in some circles – such as Mircea Eliade, spoke of the Indian time concept as the myth of the eternal return of cycles of time, precluding history.

All this apart, time was essential to the creation of cosmology and eschatology as much as a calendar was essential to historical chronology. The existence of a historical chronology and a sense of history, which some of us are now arguing are evident in certain early Indian texts, implies that there were in fact at least two concepts of time: the cyclic, found more often in the construction of cosmology; and the linear, which becomes apparent from the sources that the early Indian tradition claimed were relating the past.

I would like to argue that not only were there distinct concepts of time such as the cyclic and the linear but that these were not parallel and unrelated. I would like to argue further that there was a sensitivity to the function of each and a mutual enrichment of thought whenever there was an intersection of the two. My attempt will be to illustrate this by describing the use of both cyclic and linear time in early India, often simultaneous but arising from diverse perceptions and intended for variant purposes.

Sometimes these forms intersected in ways that enhance the meaning of both. My perspective as a historian is to view the forms and the intersections through texts associated with perceptions of the past.

The measurement of time

Concepts of time tend to be influenced by the measurement of time. A terrestrial form of reckoning was culled from the changing seasons and the diversity they brought to the landscape. The heroes of the Kuru clan mentioned in Indian texts of the first millennium BC, set out on their cattle raids in the dewy season, returning with captured herds just prior to the start of the rains. Time reckoning by seasons also encouraged what might be called ritual time. Domestic rituals were personal and focused on rites of passage, but elaborate seasonal rituals attracted a large number of kinsmen and clan members. The sacrificial altar was sometimes said to symbolise time and the ritual marked regeneration through time.

Parallel to these forms of time reckoning, a more precise measurement involved turning heavenward and was constructed on observations of the two most visible heavenly bodies, the sun and the moon, and the constellations. By the mid first millennium BC such observations provided: the scale of the lunar day, the *tithi*, with its multiple sub-divisions, the *muhurta*; the fortnights of the waxing and waning moon, the *paksha*; and the lunar month, the *masa*. But the longer periods of the two solstices – the *uttarayana* and the *dakshinayana* – were based on the course of the sun. The interweaving of lunar and solar calendars is reflected in the calculations that to this day determine the dates for most festivals.

Some changes in later centuries grew out of the Indian interaction with Hellenistic astronomy. Indian and Hellenistic kingdoms were contiguous in the north-west of the Indian sub-continent and maritime trading connections linked the western coast of India with ports along the Red Sea and the eastern Mediterranean, a link that also provided knowledge derived from navigational information. Alexandria was the location of activity in these matters. Studies on astronomy and mathematics were translated from Greek into Sanskrit.

The Indian astronomer Varahamihira of the mid first millennium AD remarked that the Yavanas – the Hellenistic Greeks and others from west

Asia – although socially of low status and outside the pale of caste society, were nevertheless to be respected as seers (*rishis*) because of their knowledge of both astronomy and astrology. Interestingly, a couple of centuries later Indian scholars resident at the court of Harun-al-Rashid in Baghdad took the knowledge of mathematics and astronomy as developed in India to the Arabs, the most widely quoted examples being Indian numerals and the concept of the zero. By this time Indian astronomy was increasingly incorporating planetary motions and solar reckoning.

Cycles within cycles

A measurement of time large enough to reflect these changes was the adaptation of the idea of the *yuga*. This was initially a five-year cycle but was gradually extended to immensely bigger spans. The word comes from the verb 'to yoke' and refers to planetary bodies in conjunction. The *yuga* was to become the unit of cosmological and cyclic time. Those projecting cyclic time measured the cycle in enormous figures, perhaps anxious to overawe their audience.

By far the largest of these was the *kalpa*, infinite and immeasurable, the period that begins with creation and continues until the ultimate cataclysmic destruction of the world. And how was this perceived? Some represented the *kalpa* spatially and these descriptions are such that they cannot be measured in temporal terms. Interestingly, they often come from sources associated with those who were regarded as heretics by the brahman orthodoxy. One Buddhist text measuring the *kalpa* says: if there is a mountain in the shape of a cube, measuring approximately 5 kilometres on each side, and if every hundred years the mountain is brushed with a silk scarf held, according to some in the beak of an eagle which flew over the mountain, then the time taken for the mountain to be eroded is a *kalpa*. The description in a text of the Ajivika sect is equally exaggerated: if there is a river which is 117 649 times the size of the Ganges, and if every hundred years one grain of sand is removed from its bed, then the time required for the removal of all the sand would be one measure of time and it takes 3000 of these measures to make one *kalpa*.

The recurring refrain of 'every hundred years' introduces a temporal dimension of humanly manageable real time, but the image is essentially spatial. It indicates the impossibility of measuring such a length of time almost to the point of negating time. The length of the *kalpa* is a deliberate transgression of time and was thought up by those who were aware of historical time. At a literal level, the silk scarf would have quickly disintegrated. And who could remove the grains of sand from the bed of a flowing river?

Time as an infinity was, however, not the view of contemporary astronomers, who did suggest a temporal length for the *kalpa*. This was 4320 million years, a figure that was large enough for their calculations. The same figure was used in cosmology, in the theory that time should be measured in the great cycles – *mahayugas*. This was one of the theories of cyclic time familiar from brahmanical texts. There is therefore an interface between cosmology and astronomy in terms of the figures used for the length of the ages and the cycles. It remains unclear whether the astronomers borrowed the figures from the creators of cosmology or vice versa. Perhaps cosmology was seeking legitimacy by borrowing the numbers used by the astronomers. A divergence between the two becomes apparent in the figures used by later astronomers that sometimes differed from these.

Each *mahayuga* or great cycle incorporated four lesser cycles, the *yugas*, but these were not of equal length. The pattern in which the great cycle is set out, and which holds together the cyclic theory, does hint at some controlling agency. One theory did maintain that time regulates the universe. The four ages were perceived in the following order: the first was the Krita or the Satya, consisting of 4000 divine years sandwiched between two twilight periods of 400 years each; then came the Treta, of 3000 years with two similar preceding and subsequent twilight periods each of 300 years; this was followed by the Dvapara of 2000 years with a twilight at each end of 200 years; and finally the Kali of 1000 years with similar twilight periods of 100 years each. These add up to 12 000 divine years and have to be multiplied by 360 to arrive at the figure for human years. A great cycle therefore extends over 4 320 000 human years.

The play is on the number 432 and it increases by adding zeros. Did this fantasy on numbers arise from the excitement of having discovered the uses of the zero at around this time? The notion of cycles may have been reinforced by the notion of the recurring rebirth of the soul – *karma* and *samsara* – which was a common belief among many religious sects. The names of the four ages were taken from the throws at dice, thus interjecting an element of chance into time. The present Kali age is the age of the losing throw. The start of the Kaliyuga was calculated to a date equivalent to 3102 BC. Since it has a length of 432 000 human years and only 5000 have been completed, we have an immense future of declining norms before us, until the cataclysmic end. By way of scale, we are also told that the length of a human life is that of a dew-drop on the tip of a blade of grass at sunrise.

The descending arithmetic progression in the length of the four cycles suggests that there was an attempt at an orderly system of numbers. Some numbers were regarded as magical such as 7, 12, and even 432, which have parallels in other contemporary cultures. The cycles are not identical and therefore permit of new events. Because of the difference in length there could not have been a complete repetition of events. It is thus possible for an event to be unique. The circle does not return to the beginning but moves into the next and smaller one. Such a continuity of circles could be stretched to a spiral, a wave or even perhaps a not very straight line. The question could be asked as to whether these should be seen as cycles or as ages? Where time is seen as a completed cycle it registers up-swings and down-swings within the cycle that are often linked to an increase or decrease in social well-being, as is suggested in the Buddhist *kalachakra* – the wheel of time. The return to the golden age would require the termination of the cycle.

The decrease in the length of each age was not just an attempt to follow a mathematical pattern. It is also said that there is a corresponding decline of *dharma* – the social, ethical and sacred ordering of society as formulated by the highest caste, that of the brahmans. The first and largest *yuga* encapsulated the golden age at its start, but subsequently there is a gradual decline in each age, culminating in the degeneration characteristic of the present Kali age. The symbols of decline are easily recognisable: marriage becomes necessary to human procreation, for men and women are no longer born as adult couples; the height and form of the human body begins to grow smaller; the length of life decreases dramatically; and labour becomes necessary. There is also an abundance of heretics and unrighteous people. These are familiar characteristics of an age of decline in the time concepts of many cultures. The decrease of *dharma* is compared to a bull that stands on four legs in the first age, but drops one leg in each subsequent age. There is a substantial change from one age to the next.

The decline inherent in the Kali age is also underlined by the description of the caste order governing social norms being gradually inverted. The lower castes will take over the status and functions of the upper castes, even to the extent of performing rituals to which they were not previously entitled. This is in part prophecy but is also a fear of current changing conditions challenging the norms. Thus kings, who are not of the *kshatriya* or aristocratic caste (see later), but of obscure origin, and frequently low caste *shudras*, or from outside the pale of caste society, can easily adopt the higher status. They are referred to as degenerate *kshatriyas* but this does not erode their authority. An even bigger disaster will be that women will begin to be liberated. This would also tie into the undermining of caste society, since the subordination of women was essential to its continuance. It shall indeed be a world turned upside down.

When the condition of decline is acute then the faithful will flee to the hills and await the coming of the brahman Kalkin, who is said to be the tenth incarnation of the deity Vishnu, and who will restore the norms of caste society. Kalkin is a parallel concept to that of the coming of the late Buddha, the Buddha Maitreya, who will save the true doctrine from extinction and re-establish Buddhism. It is interesting that many of these saviour-figures either emerge or receive added attention around the early Christian era, when the belief systems to which they belong – Vaishnavism, Buddhism, Zoroastrianism and Christianity – are in close contact in the area stretching from India to central Asia and the eastern Mediterranean. The coming of Kalkin (Figure 1) can be read as an alternative to the cataclysmic end of the *mahayuga*, since he initiates another golden age. History does not end and time does not cease, but the eschatology is perhaps implicit, given the immense length of the great cycle.

The *Vishnu Purana* and categories of linear time

The Kaliyuga was a concept frequently referred to in a variety of sources, but the details of the cyclic theory come in particular texts. Among these were: the long epic poem, the *Mahabharata*, initially composed in the first millennium BC; the code of social duty and ritual requirements known as the Manu *Dharmashastra*, written at the turn of the Christian era; and the more accessible and popular religious texts of the early centuries AD, the *Puranas*. The inclusion of theories of cyclic time in the epic are in the sections

FIGURE 1. A 20th century decoration on the walls of the palace at Dhrangadhra (Gujerat), representing the coming of the god Vishnu in the Kali age in the form of Kalkin, the saviour figure, riding a white horse.

generally believed to be later interpolations and thought to have been inserted by brahman redactors when the epic was converted to sacred literature. The authorship of the *Dharmashastras* was also brahman. Although many of the *Puranas* are said to have been composed by bards, in effect they are again largely edited by brahman authors. There is therefore a common authorship supporting these ideas.

The historiographical link with modern theories is that these were the texts studied and translated by Orientalists such as William Jones and H. H. Wilson. These studies were encouraged, with the intention of enhancing British understanding of pre-colonial laws, religious beliefs and practices and in searching for the Indian past. Knowledge about the colony was described as the necessary furniture of empire. But because these particular texts were given importance initially, their description of cyclic time came to be seen as the sole form of time reckoning in India. One can understand how a utilitarian such as James Mill dismissed Indian concepts of time as pretensions to remote antiquity, but it is more difficult to explain why H. H. Wilson did not recognise the linear pattern of time in, for example, the *Vishnu Purana*, a text on which he worked at length and which he translated.

In relating the details of what happened in the Kaliyuga the *Vishnu Purana* provides us with various categories of linear time. The *vamsha-anu-charita* section of the text consists of genealogies and descent lists of dynasties. The genealogies are of the chiefs of clans, referred to as *kshatriyas* and they cover about 100 generations. They need not be taken as factual records but can be analysed as perceptions of the past. The word used for the descent group is *vamsha*, the name of the bamboo and an obviously appropriate symbol, since the plant grows segment by segment, each out of a node. The analogy with genealogical descent is close. The imagery emphasises linearity, which is expressed in what might be called 'generational time', seen as the flow of generations (Plate I). This construction of the past dates to the early centuries AD, and was subsequently used to manipulate the claims and statuses of later rulers through a variety of assumed links.

But the flow is not unbroken. There are time-markers separating categories of generational time. The first time-marker is the great Flood, which enveloped the world and which separates the pre-genealogical period from the succession of generations of clan chiefs. Each of the rulers of the antediluvian period ruled for many thousands of years. At the time of the Flood, the god Vishnu in his incarnation as a fish appears to the ruler Manu, and instructs him to build a boat. This is tied to the horn of the god-fish, is towed through the flood waters and lodged safely on Mount Meru. When the Flood subsides, Manu emerges from the boat and becomes the progenitor of those who are born as the ruling clans. The Flood is first mentioned in a text of about the eighth century BC and is later elaborated upon in the *Puranas*. It has such close parallels with the Mesopotamian legend that it may well have been an adaptation.

Subsequent to the Flood the supposed genealogies of the ancient heroes or the *kshatriyas* are mapped. The succession of generations is divided into two groups named after the sun and the moon, a symbolism of both dichotomy and eternity, used frequently in myth, yoga, alchemy and on many other occasions. The solar and the lunar lines mark a different pattern of descent. The solar line or the Suryavamsha emphasises primogeniture and claims to record the descent only of the eldest sons. The pattern of descent therefore forms vertical parallels. In the epic the *Ramayana*, the families of status are of the solar line. The lunar line or the Chandravamsha is laid out in the form of a segmentary system and the lines of descent fan out, since all the presumed sons and their sons are located in the system. The advantage of a segmentary system, or one similar to it, is that it can more easily incorporate a variety of groups into a genealogy by latching them onto the existing ones. These constitute the structure of society in the other epic, the *Mahabharata*.

The solar line slowly peters out. But those belonging to the lunar line are brought together in the second time-marker, the famous war said to have been fought on the battlefield at Kurukshetra near Delhi, and described in the *Mahabharata*. Virtually every hero of that period was involved in the great battle. Many are not heard of after the event. The war we are told terminated the glory of the ancient heroes and the *kshatriya* aristocracy. In the representation of the past, the war demarcates the age of heroes from that which followed. This was the age of dynasties. A major indicator of change is that the narrative switches from the past tense to the future tense and reads as a prophecy. This also invokes astrology, especially popular in court circles.

Genealogies incorporating generational time, I would argue, are within the framework of linear time. The texts included in what is called the ancient Indian historical tradition – the *itihasa purana* – make claims to representing the past 'as it was'. The Flood seems to demarcate the time of myth from the time of history. There is a distinct beginning from after the Flood and an equally distinct termination in the war. The arrow of time moves steadily through the generations and to the battlefield. That the lists may have much that is fabricated – as is the case with all such lists – is not so relevant as is the perception of the form of time, which is linear. This is further underlined in the next section of the *Vishnu Purana* recording the dynasties ruling over a major part of northern India.

The narrative of the dynasties in this section of the *Vishnu Purana* is limited largely to the names of rulers, with an occasional but minimum commentary. Regnal years are sometimes included, further highlighting a sense of linear time. The dynasties, unlike the *kshatriya* families of the earlier section, have no kinship ties with each other and were rarely even of the *kshatriya* caste to which rulers are supposed to belong. In practice the profession of ruling seems to have been open to any caste, which was a reversal of the norm and was another example of the reversals characteristic of the Kaliyuga. The names of dynasties and rulers are sometimes corroborated in other sources such as inscriptions, which were at that time being issued in large numbers.

The creation of eras

Thus the section in the *Puranas* describing the succession of those who ruled encapsulates three kinds of time. The prediluvian rulers, the Manus, are referred to in what could be called cosmological time, beyond even the purview of the great cycles. This is almost a form of reaching back to time before time. It is distant from the two more human time frames: genealogies and dynasties. With these, the presence of what is conventionally regarded as history begins to surface. This move in the direction of historical time may have been associated with another form of measuring time more closely linked to history, namely the creation of eras.

The use of a particular era, the *samvatsara*, related to historical chronology, probably grew out of a consciousness of enhanced political power, with a focus on the royal court. The earliest inscriptions, those of the Mauryan emperor Ashoka ruling in the third century BC, are dated in regnal years counted from his accession. This may have provided an impetus for establishing an era that would be a commonly accepted base point for historical dates. The start of the earliest era was the much used Krita era of 58 BC, later to be called the Malava era but more popularly known as the Vikrama era. There has been considerable controversy regarding its origin. The current consensus associates it with a relatively unimportant king, Azes I. Its impressive continuity to the present suggests associations other than just the accession of a minor king, for eras are often abandoned when a dynasty declines. There might have been a connection with astronomy, since the city of Ujjain, an important focus for various calculations in astronomy, was located in the territory claimed by the Malavas.

Historical events do, however, become the rationale for starting eras subsequently, such as the Shaka era of AD 78, the Chedi era of 248–249, the Gupta era of 319–320, the Harsha era of 606, and so on – a virtual blossoming of eras, most of which were commemorating accession to kingship. Many of those who started these eras were, in origin, small-time rulers who had succeeded in establishing large kingdoms. As a status symbol, the Chalukya-Vikrama era of AD 1075 was not only a claim to supremacy by the Chalukya king Vikramaditya VI, but included the justification for his usurping the throne. The creating and abandoning of eras became an act of political choice. The continuity of an era is not just the continuity of a calendar but also of the associations linked to what the era commemorates. The ideology implicit in starting and sustaining an era calls for historical attention.

Events related to dynastic history were not the only occasion for starting an era. Time reckoning based on the year of the death of the Buddha, the *maha-pari-nirvana*, became current in the Buddhist world. The date generally used was 486 or 483 BC. Recently, however, some scholars have questioned these dates and would prefer to bring the date forward by anything up to a century. Nevertheless, what is important is that events described in Buddhist texts are generally dated from the death of the Buddha, which is calculated as a definitive date within a particular Buddhist school.

Buddhist chronicles demonstrate a concern with time and history in that they record and narrate what they regarded as historically important events: for example, the history of the Buddhist Order or Sangha starting with the historical founder Gautama Buddha (Figure 2); relations between the Buddhist Order and the state; the founding of breakaway sects and the events that led up to these; records of gifts of land, property and investments; and the founding of monasteries and matters of monastic discipline. All these are tied to linear time in various ways. The Buddhist calendar was pegged to what were viewed as events in the life of the Buddha and the history of the Order. The linear basis of this chronology was nevertheless juxtaposed

FIGURE 2. The Rummindei Inscription marking the birthplace of the Buddha. The pillar with the inscription was put up at the order of the emperor Ashoka of the Mauryan dynasty in the third century BC.

with ideas of time cycles, as for example in the concept of the *kalachakra* or wheel of time. These had their own complexities distinct from those of the *Puranas*. This was not specific only to Buddhism. Jaina centres from the first millennium AD maintained the same kind of records. This involved histories that, in order to be legitimate, had to cohere up to a point. Such histories were not always intended to be taken literally, and certainly cannot be so taken today. They have to be decoded through the social and cultural idioms prevalent when they were written.

Dynastic chronicles and regional histories

Historical time is a requirement for what have come to be regarded as the annals of early Indian history. These are inscriptions issued by a variety of rulers, officials and others. They frequently narrate, even if briefly, the chronological and sequential history of a dynasty. Some were legal documents conferring rights in land and were proof of a title deed. Precision in dating gave greater authority and authenticity to a document. The granting of land or property to religious beneficiaries had to be made at an auspicious moment so as to carry the maximum merit for the donor. The auspicious moment was calculated by the astrologer in meticulous detail and was mentioned in the inscription recording the grant. Other categories of grants also carried precise dates. It is this precision that enables us to calculate the dates of the inscriptions in the equivalent date of the Gregorian calendar. Much of early Indian historical chronology is founded on the calculation of these dates carefully worked out by Indologists. Yet, curiously, these scholars made little effort to go beyond the bare bones of chronology and deduce the time concepts reflected in these dating systems.

Inscriptions recorded the official version of the events of a reign and were issued by almost every ruling family. The legitimising of power, especially in a competitive situation, included a range of activities. Among them was the making of grants of land, particularly to religious beneficiaries who would then act as a network of support for the ruling family. This was the occasion for obscure families who had risen to rulership to claim a status equivalent to that of established ruling families, a claim which the beneficiaries were ready to substantiate. The document accompanying the grant had to be inscribed on imperishable material – copper or stone (Figure 3). Grants had to be impressive and often more generous than those of earlier times or of competing rulers.

From about the seventh century onwards there is an efflorescence of another category of historical texts that combine these elements of linear time such as genealogies, dynastic histories and eras. These were biographies of kings or an occasional minister – the *charita* literature. The subject of the biography was a contemporary ruler, and the biography narrated the origins of his family and the history of his ancestors, particularly that which led the family to power. The central event of his reign, as assessed by the biographer and presumably the king as well, was described with appropriate literary elaboration, sometimes quite flamboyantly, and frankly eulogistic as is frequent in courtly literature. Often the intention was to defend the usurping of a throne and overturning the rule of primogeniture. Sometimes the intervention of a deity was required to justify the action of the king. And if the interventions became too frequent, the reader would understand that their intention was other than what was being related. Whatever the inten-

FIGURE 3. Four sets of copper plate inscriptions held by rings and seals, as found inside an earthenware pot from Andhavaram.

tions of the biographies, they did describe and present some significant events of a king's reign in a linear succession.

Dynastic chronicles and regional histories also drew legitimacy from linear time. These were the *vamshavalis*, literally the path of succession. The most famous among these was the much quoted *Rajatarangini* of Kalhana, but similar, although less impressive, narratives come from other regions. At the point when a region changed from being viewed as the territory of a chiefdom and came to be seen as the state claimed by a dynasty, the records of the past were collated, and a chronicle was put together. This was maintained as an up-to-date narrative of what were regarded as significant events. In their earlier sections, such chronicles incorporate some of the genealogies of the ancient heroes of the *Puranas* to whom they link the local rulers. Writing the chronicle of a region became another form of recognising the region as an entity and legitimising its succession of rulers.

In these texts time is linear and the assumptions of cyclic time may be implicit but remain distant. Cyclic time is not denied and is present in the larger reckoning. Deities and incarnations tend to be placed in the earlier cycles. But events relating to the human scale are more properly expressed as part of linear time, which was the more functional (Figure 4). This did not preclude a reference on occasion to cyclic time. A seventh century inscription records an event in the Shaka era of AD 78 and includes for good measure a reference to the date of the Kaliyuga. Whereas a reference to something like the Kaliyuga might be added, what seems to have been required was a historical date. Linear time is included within the Kaliyuga and this creates an intersection between the two forms. It is almost as if a segment of the cycle is stretched to a more linear form.

Too great an insistence on the negative characteristics of the Kaliyuga may in any case not have been complimentary to the kings who were subjects of the biographies and the chronicles. This was anyway the time of the losing throw and the ebbing away of *dharma*, but the contemporary present could hardly be described as a period of decline and retrogression if the biography was intended as a eulogy. A longer continuity of time is assumed in the inscriptions, one that went beyond even the great cycle, for the formulaic phrase always reads that the grant should last 'as long as the moon and sun endure'. Time was thought of at many levels.

Conclusion

My intention has been to suggest that various forms of time reckoning were used in early India and that concepts of both linear and cyclic time were familiar. The choice was determined by the function of the particular form

FIGURE 4. The Aihole Inscription of Pulakeshin, AD 634–635. The inscription is a poem by a certain Ravikirti and gives a eulogistic account of the exploits of Pulakeshin II.

of time and those involved in using it. Sometimes the forms intersect and at other times the one encompasses the other. The *Vishnu Purana* in one section describes at length the various ages of cyclic time. In another section it provides details of the genealogies of the heroes and of the rulers of various dynasties in the Kali age.

As part of history, time was tied into social and political functions and these can be seen either in the diverse authorship of some historical traditions or else in the compositions of the authors from the same social group using time for diverse purposes. It is stated that genealogies were originally compiled by the bards, and presumably incorporated the world-view of the ruling clans and were intended to record their past. But when this compilation came to be edited by the brahmans, in order to use it for establishing the legitimacy and social claims of their patrons who were often upstart rulers, this past appears to have been taken away from the bards and appropriated by the brahman authors of the *Puranas*. The cyclic concept was comprehensive, although distant, and was more apposite to the ritual and other concerns of the brahman priests and their perceptions of the past, requiring as it did an emphasis on that which is beyond human control. The cyclic concept became a temporal frame within which any past could be incorporated. These two concepts of time therefore are also related to the particular interests of two socially distinct groups. This is not to deny that other theories also went into the making of these concepts, or that these concepts are present in other religious ideologies, but rather to observe them from a historical perspective.

The past was sought to be captured. One way of doing this was to associate it with various projections of time. As the creators of cosmologies, the brahmans often refer to time in the cyclic form of the four ages. As keepers of the genealogies or composers of inscriptions or authors of royal biographies, the immediate point of reference is linear time. In spite of this intersection or encompassment, the function of each form is differentiated. The simultaneous use of more than a single form and its layered representation indicates some awareness that different segments of a society may view their past differently. For the historian to recognise this requires a certain sensitivity in seeing the past as multiple perceptions within the intricacies of the use of time.

The presence of more than one form of time in the same text is perhaps intended to point us towards different statements being made about each. Within linear time there can also be differentiation. Genealogical time based on a succession of generations is always at the start of the record and precedes that which we would recognise as conventional history. This is evident from the succession lists in the *Puranas*, as also from the regional chronicles. This format underlines continuity. But it is also a way of differentiating two categories of the past with the deliberate and consistent placing of one before the other.

After the mid first millennium AD, the past, where feasible, tended to be introduced into the construction of ancestry and claims to legitimacy, and in rights to property. This was likely to be more so where claims were being contested. The past involved multiple views of time. For many, the fourth age, although part of the great cycle or *mahayuga*, encapsulated nevertheless the linear forms of the perceived history of heroes and kings. Eras became fashionable and necessary, precise dating systems came to be used in the epigraphic annals of the various dynasties, and regional societies were poised to patronise the writing of royal biographies and the chronicled histories of the past. A sense of history was perhaps embedded in some sources but was visible in others.

The insistence on Indian society having only a cyclic concept of time may not be the general view any longer. But even its rejection has not yet encouraged the recognition of forms of history as evident in some early Indian texts. Such recognition is likely to be strengthened through a demonstration of the presence of linear time. Given that every society has an awareness of its past, it is perhaps futile to construct a society that denies history only in order to argue that it is unique or different from what is believed to be the norm.

The two time concepts do not exhaust the variations on time that in Indian texts are portrayed in diverse images. Some maintain that time was the creator, begetting the sky and the earth, the waters and the sun, the sacrifice and the ritual verses: it drove a horse with seven reins, was thousand-eyed and ageless. Or it was the imperishable deity through whom everything that has life eventually dies. For others, time was the ultimate cause lying between heaven and earth and weaving the past, present and future across space. Equally evocative is the projection of time as that which directs and regulates the universe – the *sutradhara*.

The creators of myths, the chroniclers of kings, the collectors of taxes and those who pay them, subscribe to divergent concepts of time. Distinctions can be made between cosmological time and historical time, but the degree of separation or of overlap would vary according to historical situations and the way in which they are perceived. The first could be a fantasy on time, although a conscious fantasy, carefully constructed and therefore reflective of its authors and their mythologies. The second draws on functions of a more measured time, also carefully constructed but reflecting more manageable concerns. If time is to be seen as a metaphor of history, which is what I have been suggesting, then perhaps we need to explore the many more patterns of time and their historical intersections.

FURTHER READING

Eliade, M., *Cosmos and History: The Myth of the Eternal Return*, Princeton, NJ: Pantheon Books, 1954.

Eliade, M., 'Time and eternity in Indian thought', in *Man and Time*, Papers from the Eranos Yearbooks, pp. 173–200 (Bollingen Series, XXX.3), New York, Pantheon Books, 1957.

Pathak, V. S., *Ancient Historians of India*, Bombay: Asia Publishing House, 1966.

Pingree, D., *Jyotihsastra*, Wiesbaden: Otto Harrassowitz, 1981.

Thapar, R., *Interpreting Early India*, Delhi: Oxford University Press, 1992.

Thapar, R., *Time as a Metaphor of History: Early India*, Delhi: Oxford University Press, 1996.

3 Time Travel

D. H. MELLOR

The passing of time

One of the difficulties of talking sense about time travel is that it means different things to different people. For some it means the ineluctable passing of time, while for others it means the exotic activities of time travellers such as Dr Who. Yet, although these are different, they are not unrelated, and to talk sense about the latter I must first say something about the former.

The sense in which we all travel in time is the sense in which time passes, as it always has and always will. If that is time travel, there is no doubt that it occurs, and occurs automatically. In this sense we have no choice but to travel in time: it is not something we can choose to do, more or less easily. It just happens to us, as to everything else, whether we like it or not.

The main problem posed by time passing is how to make sense of it passing more or less quickly, as it often seems to do. The best way to see the problem is to compare the rate at which it passes with the rates at which other changes occur, as when space passes by a train taking an hour to cover the 60 miles from London to Cambridge. During that journey, space is passing at a mile a minute, a rate that is both objective and variable, since the train could go either faster or slower. Similarly for other changes, such as changes in size or temperature: things can get larger or smaller, or warm up or cool down, at different rates, just as we can travel through space at different rates.

But we cannot in this sense travel through time at different rates. It may take more or less than an hour to go from London to Cambridge, but how can it take more or less than an hour to go from 10 a.m. to 11 a.m.? That it takes 60 minutes to get through an hour, far from being a variable matter of fact, as it is with other changes, is a trivial tautology, on a par with having to travel a mile to reach to a place that is a mile away.

How then *can* time pass more or less quickly? Well, imagine a lecture that in fact lasts an hour but seems to last much longer, say two hours, so that during the lecture time seems to pass at a rate of two hours per hour. That is the sort of thing we mean when we talk of time passing slowly. There is nothing mysterious about it; nor need this sense, in which time *can* pass slowly, be purely subjective. By this I mean not just that a lecture can be equally boring for everyone, but rather that the experience of time passing slowly can and presumably does have a physiological explanation. Take the experience, which many people have had, of facing a sudden and unexpected danger, such as an imminent car crash. Suddenly everything seems to be happening – i.e. time seems to be passing – very slowly, apparently because, for obvious reasons, the adrenaline released in our bodies in such emergencies speeds up our mental processes, including those that give us our sense of how fast time is passing.

In other words, when time seems to be passing slowly what is really happening is that, as measured by external clocks, our internal clocks have speeded up. Similarly when time seems to be passing quickly, as when a fascinating lecture ends all too soon, the hour seems to have flown by in only half the time, i.e. at an apparent rate of 30 minutes an hour. This too is a familiar phenomenon, and again the explanation is the same: a mismatch between internal and external measures of the time interval between two events, in this case the beginning and end of a lecture. Only here the mismatch goes the other way: as measured by external clocks, our internal clocks have slowed down.

So much for time passing more or less quickly, a phenomenon that is neither uncommon nor problematic. In particular, the fact that one clock may speed up or slow down relative to another in no way conflicts with the tautological fact that, by any single clock, it must take 60 minutes to get from 10 a.m. to 11 a.m. as measured by that clock.

Forward time travel

The real interest of time passing is what it tells us about forward time travel, namely that there must be more to it than time passing at 60 minutes an hour. For forward-travelling time machines to be worth having, they must take us into the future faster than that. And what the fact that time can pass quickly shows is that this *is* possible in principle, the only question being how to do it in practice. But first we must see more clearly what doing it does and does not entail.

Suppose we hire Dr Who's time machine *TARDIS* to take us 100 years into the future – say from 2000 to 2100 – in an hour. What must it do to do this? Well, one thing it need not do is what Dr Who programmes always make it do, namely disappear *en route*; that is, be *nowhere* in space between leaving 2000 at 10 a.m. and arriving in 2100 an hour later at 11 a.m. It need no more do that than a train travelling from London to Cambridge needs to vanish in between. On the contrary, since vanishing in between may well be impossible, requiring trains to do it could make ordinary travel through space seem impossible when we know it is not. But if ordinary train travel is not to be falsely ruled out by requiring trains to vanish *en route*, we should not risk ruling out forward time travel by making time machines vanish *en route*. To give *TARDIS*'s one hour trip from 2000 to 2100 a decent prospect of possibility, we must allow it to be somewhere at all times in between.

This illustrates a maxim, about travel of all kinds, which I shall need again later and should therefore make explicit now. The maxim is that, even if it is sometimes better to travel hopefully than to arrive, *if you do not arrive somewhere you have not travelled there*; *if you do arrive, you* **have** *travelled there, regardless of how you did it*. We shall see later how this maxim affects the prospects of backward time travel; what matters here is that it helps those of forward time travel by removing the needless impediment of in-flight disappearance.

What then *does* forward time travel require? Consider what needs to happen outside and inside *TARDIS* between its setting off in 2000 and its arriving an hour later in 2100. Outside *TARDIS* a century passes between these two events: all clocks and calendars move on 100 years; 100 birthdays pass; 100 sets of seasons come and go; and so on. In short, all cyclic and one-way processes outside *TARDIS* reach points that they normally take 100 years to reach, whereas inside *TARDIS* the same processes reach points they normally reach in only 60 minutes. No seasons come and go, the clocks move on only an hour, and far from ageing even a day, let alone a century, our time travellers hardly have time to digest the coffee and biscuits served after take-off. This is what it takes – and all it takes – to travel forward 100 years in an hour.

To see what this amounts to, consider how fast events outside *TARDIS* must seem to be happening to those inside (and vice versa). To those inside, the outside world will look like an amazingly fast-forwarding video, with 100 years of events happening in an hour. To those outside, everything inside will seem to be happening incredibly slowly, with clocks taking 100 years to move on an hour, people taking a year to utter a sentence, a decade to drink a cup of coffee, and so on. In other words, in order to travel forward in time, all *TARDIS* needs to do is to make all processes within it run very slowly by comparison with the same processes in the world outside. How can that be done?

There are at present two known ways of doing it: a high-tech way and a low-tech way. The high-tech way exploits Einstein's special theory of relativity, as follows. Suppose a spaceship leaves the earth in 2000 and returns in 2100, travelling out and back very fast indeed. It follows from Einstein's theory that, on its return, the spaceship and all its contents will have aged less than they would have done on earth. Moreover, the faster the spaceship travels, the less it and its contents will have aged, so that if it travels fast enough – that is, close enough to the speed of light – they will have aged only an hour. In that case it will have taken them just one hour of their time to go from the earthly events of 2000 to those of 2100; if that is not forward time travel, what is?

However, this method of time travel, while certainly possible in principle, is at present far too costly to be feasible in practice. Its low-tech alternative, on the other hand, is not only feasible but commonplace. This alternative relies on the fact that most chemical (and hence most biological and psychological) processes slow down when it gets colder, with their rates roughly halving with each 10 degrees Celsius drop in temperature. Hence hibernation, where animals, by lowering their body temperatures, slow down their metabolic processes to reduce their need for energy and hence food when, as in hard winters, food is hard to get – hence cryonics, which can stop people ageing for years by freezing them. But this, as we have seen, is all it takes to produce forward time travel. So when we can freeze and unfreeze people without killing them, cryonics will provide a perfect method of time travelling, by freezing people when they want to set off and defrosting them, unaged, at whatever time they wanted to arrive.

Meanwhile, it follows from all this that domestic freezers and refrigerators, whose job is after all to slow down processes of ageing and decay in their contents, qualify as forward-travelling time machines! That, however, is neither an impressive fact about freezers nor a refutation of this concept of forward time travel. All it shows is that, conceptually, forward time travel is no big deal, being in reality nothing but slow ageing.

Backward time travel

If forward time travel is slow ageing, backward time travel is not fast ageing but something quite different. Suppose for example that *TARDIS* sets off from Cambridge in 2050 and, an hour later by *TARDIS* time, arrives in London in 1950. This is backward time travel, and what makes it so is the fact that, by *TARDIS* time, *TARDIS* leaves an hour *before* it arrives, while in outside time it leaves a century *after* it arrives. This reversal of the external time order of events is what makes backward time travel quite different from its forward counterpart, as we can see by noting that the difference in time *span* is immaterial. For while taking a century to go from 2000 to 2100 is not forward time travel but merely time passing, taking a century to go from 2050 to 1950 is still backward time travel, however slow. We cannot assume therefore that backward time travel will be as unproblematic as forward time travel; indeed it is not, as we shall see.

One objection to backward time travel, however, we may dismiss at once. This is that two events cannot have opposite time orders, since nothing can be both earlier and later than something else, any more than it can be both hotter and colder than something else. That this objection is too quick we can see by considering that it would also rule out forward time travel, since it seems equally true that nothing can be both *one hour* and *a century* later than something else. But it can, since different spacetime routes between two events can easily have different temporal lengths, just as different spatial routes between two places can have different spatial lengths. There is therefore no contradiction in saying that routes outside *TARDIS* between event *d* (*TARDIS*'s departure) and event *a* (*TARDIS*'s arrival) are a century long, while routes inside it are only an hour long. But then it is not obviously impossible for *d* to be earlier than *a* on routes linking these events inside *TARDIS* and later than *a* on routes linking them outside it.

To see what, if anything, makes backward time travel impossible, we must first recall the maxim on p. 48, that to travel to somewhere you must arrive there. Thus, imagining that *TARDIS will* travel from Cambridge in 2050 to London in 1950 must include imagining that it *did* arrive in London in 1950. And by this I mean imagining that the London it arrived at in 1950 was *our* London, not some London look-alike in another parallel or merely possible world. This is because arriving in another London is just a way of *not* arriving in the actual one, and hence, given our maxim, of not having travelled there. In short, therefore, and in general, if a time machine ever *will* arrive in our past, then it *did* arrive in our past.

Next we must ask what it takes to arrive in London in 1950, i.e. what it takes to *be* there then. First, in the only sense of 'being there' that matters here, it is obvious that we cannot be in London in 1950 just by *thinking* about London then. If we could, time travel would be trivially easy, since no one denies that we can travel into the past in thought. But that is not the issue: the issue is whether we could travel there in reality, which must therefore mean more than travelling there in thought.

For the same reason, we cannot be in the past just by *reading* about it. Old documents, or well-written histories, may in some sense 'take you there', but not in the sense that matters here. Nor can pictorial representations of past events take us there in the sense we need: not paintings, not photographs, not films. Not even television or radio can do it: my seeing or hearing in Cambridge a broadcast live from London does not literally take me to London. And as with other places, so it is with other times. Looking through telescopes at celestial events millions of light years away, and hence millions of years ago, no more transports us to those remote times than it does to those remote places: telescopes are not time machines.

What then *does* it take to be somewhere in space and time? The answer is that, to *be* there, you must be able to affect, as well as be affected by, whatever else is there. In other words, you must be able to *interact* with it, just as you can interact with the copy of this book that you are now reading. It is after all only because you and this book are together in the same place *at the same time* that you can not only be affected *by* it (e.g. by seeing it) but also *affect* it (e.g. by closing it). Similarly with the other things with which this book can interact more or less directly, like the surface it is resting on, which *when* they are in contact it simultaneously weighs down and is supported by. *How* things and people can interact in this direct way depends of course on what kinds of things and people they are; but that they *can* do so is what makes them adjacent in time as well as space. In particular, then, for *TARDIS* and its passengers to be somewhere in 1950, they must be able to interact with the other things and people that were there then, just as those things and people interact with each other. Otherwise, wherever *TARDIS* travelled to, it was not 1950.

The direction of time

If this is what it takes for *TARDIS* to *be* in 1950, what makes it *arrive* as opposed to *leave* then? The answer I gave above was that, by *TARDIS* time, *TARDIS* arrives (event *a*) in 1950 *after* it leaves (event *d*) in 2050. This reversal of the external time order of those two events is what makes *a* an arrival and *d* a departure, rather than the other way round, thus making this time travel backward rather than forward. But then what gives *a* and *d* these opposite time orders? What, in other words, gives time a different direction inside and outside *TARDIS*?

Since, as we have noted, it is immaterial how long it takes to travel back in time, let us now assume that it takes a century to go back a century, an hour to go back an hour, and so on. This lets us simplify matters by comparing backward time travel not with its forward counterpart but with ordinary space travel. Look then at Figure 1, which shows the so-called 'world lines' of two non-stop trains travelling in opposite directions between London and Cambridge, leaving at 10 a.m., passing each other at the town Hitchin in between at 10:30 a.m., and arriving at 11 a.m.

Now imagine a Cambridge (i.e. London-to-Cambridge) train travelling backward in time, leaving London at 11 a.m. and getting nearer to Cambridge not at later but at earlier times, and arriving therefore at 10 a.m. Then, as this train is at all the same places at all the same times as the ordinary London train, they share the same world line, just as the ordinary Cambridge train and a time-travelling London train would do. How then does an ordinary train differ from a time-travelling one with the same world line?

The difference seems to lie in the direction of one-way processes. For example, conversations on ordinary London trains start nearer to Cambridge than they finish, while on time-travelling Cambridge trains they start nearer to London. Similarly for all other one-way processes, such as those of watches, which on London trains read '10 a.m.' in Cambridge and '11 a.m.' in London

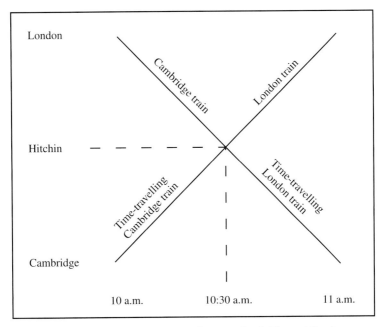

FIGURE 1. World lines of non-stop trains between Cambridge and London.

and on time-travelling Cambridge ones would read '10 a.m.' in London and '11 a.m.' in Cambridge. And as in space, so in time. By outside time, conversations in time-travelling trains stop before they start, the hands of watches go round anticlockwise, and so on. The direction of one-way processes seems then to be what, by giving time its direction, distinguishes things that are travelling backward in time from things that are not.

But what then of things, such as fundamental particles, that contain no one-way processes? Take electrons and positrons, which differ only in the sign of their electric charge, the former being negative and the latter positive. Thus, as like charges repel and unlike ones attract, negative charges repel electrons and attract positrons, while positive charges repel positrons and attract electrons. So now imagine two fixed charges (a negative one N and a positive one P) and two particles moving between them (an electron going from N to P, and a positron going from P to N). Suppose also, to maintain the parallel with our trains, that each particle leaves at 10 a.m. and arrives at 11 a.m., as shown in Figure 2.

Finally, imagine an electron travelling backward in time, leaving N at 11 a.m. and growing nearer to P not at later but at earlier times, and arriving therefore at 10 a.m. Then, as this electron is at all the same places at all the same times as our ordinary positron travelling from P to N, the electron and the positron share the same world line, just as our ordinary electron and a time-travelling positron would do. How then do ordinary electrons and positrons differ from their time-travelling opposite numbers, given that, unlike trains, they contain no one-way processes whose opposite directions could distinguish them?

The answer, according to some physicists, is that positrons and electrons do *not* differ: positrons just *are* electrons travelling back in time. If this is so, then backward time travel must be possible, since it actually occurs. But it is not so, for two reasons. First, as Figure 2 shows, we have no more reason to call positrons time-travelling electrons than to call electrons time-travelling positrons. If the time order of events on world lines is given by the direc-

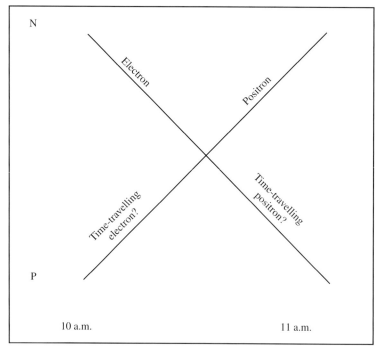

FIGURE 2. World lines of electrons and positrons.

tion of one-way processes, which neither electrons nor positrons contain, then nothing gives either of them a temporal direction: that explanation of their opposite reactions to negative and positive charges is vacuous. Setting aside the tautological passing of time, we have no reason to credit either electrons or positrons with any ability to travel in time at all.

The second reason for denying that positrons are time-travelling electrons is that, despite appearances, time does *not* get its direction from any one-way process, even though many philosophers and physicists think it does. The leading contenders for this role are the expansion of the universe, the increasing entropy of isolated systems, and the fact that light travels away from its sources rather than converging on them. But we need not go into details to see why none of these processes can do the job. Of course each of them *has* a direction, which correlates with the direction of time: if they did not, they would not be one-way processes. But it takes more than this for a one-way process to *give* time its direction. For it to do that, reversing the process's own direction must be not merely physically but *logically* impossible; for, since doing so would by definition also reverse the direction of time, the two reversals would cancel out. Thus if the universe expanding is what gives time its direction, it will have to expand for ever as a matter of logic. But whether the universe will do that depends not on logic but on contingent physical facts, such as whether it is dense enough for gravity to halt and reverse its initial expansion. But that can be the contingent matter it evidently is only if the direction of time is logically independent of that in which the universe has expanded so far.

The other two processes are even less credible sources of time's direction. For although each of them mostly goes one way, neither of them always does so. The entropy of isolated systems can decrease, and light can converge, as it does for example whenever a camera lens makes it form a photographic image. If that reversed the direction of time inside a camera, it would make the light in the camera travel backward to the lens from a previously formed image, and then meet and annihilate the light coming in from the object photographed! No one believes it does that; yet that is what is implied by identifying time's direction with that in which light diverges.

What really gives time its direction is *causation*, as we can see by asking how stopping our Cambridge to London train at Hitchin would affect its world line. The answer is of course that, while its world line would still run from Cambridge at 10 a.m. to Hitchin at 10:30, it would then remain at Hitchin from 10:30 to 11 a.m., as shown in Figure 3. Stopping our time-travelling Cambridge train at Hitchin, on the other hand, would leave its world line running from London at 11 a.m. to Hitchin at 10:30 but make it stay at Hitchin from 10:30 to 10 a.m., as Figure 3 also shows. The reason in both cases is that the effects of stopping a train at Hitchin occur, like all effects, *later* than their causes – which in a time-travelling train, where the direction of time is reversed, means *earlier* in outside time. This is what makes the actual trains whose world lines run from Cambridge at 10 a.m. to London at 11 a.m. be ordinary London trains rather than time-travelling Cambridge ones: the fact that, by outside time, all the effects of stopping or otherwise affecting them occur *after* and not *before* the causes of those effects.

Similarly with positrons and time-travelling electrons, as shown in Figure 4. Stopping our positron at 10:30 a.m. at a point H halfway between N and P would leave its world line running from P at 10 a.m. to H at 10:30 but make it stay at H from 10:30 to 11 a.m. It would *not* leave its world line

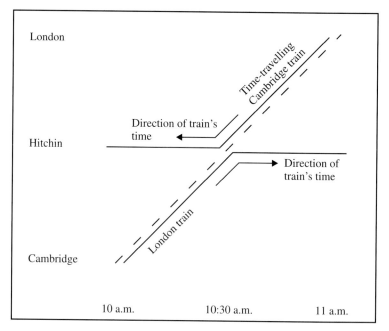

FIGURE 3. The effects of stopping at Hitchin.

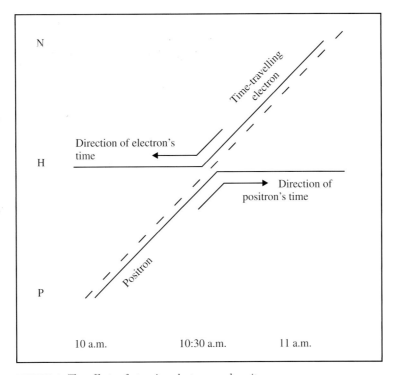

FIGURE 4. The effects of stopping electrons and positrons.

running from N at 11 a.m. to H at 10:30 while making it stay at H from 10:30 to 10 a.m., as it would if it were an electron travelling back in time. This is what makes positrons ordinary positively charged particles, and not time-travelling electrons: the fact that, by outside time, all the effects of stopping or otherwise affecting them occur *after* and not *before* the causes of those effects.

Soluble problems of time travel

Positrons and other so-called anti-particles are therefore *not* ordinary particles travelling backward in time; I know of no other credible cases of actual backward time travel. But discrediting all actual examples of a phenomenon does not show it to be impossible. To show that we must find some feature which is both essential to it and impossible. What in this case might that feature be?

Boundaries

Many objections to backward time travel fail because they rely on features that either it need not have or that, although odd, are not obviously impossible. The problems posed by boundaries between backward-travelling time machines and the outside world are a case in point. Suppose we ask how, in a time-travelling Cambridge train, I could see something – a sheep, say – out of the windows. Normally I see such things by being affected by the photons they reflect into my eyes. But here the photons have to pass from the world outside, where time runs one way, into the train, where it runs the other way, a transition of which it is hard to make good sense. For by outside time, since the train and all its contents are travelling back in time, the photon reaches my eye not after but *before* it goes through the window. In other words, as in the camera mentioned on p. 55, by outside time there appear to be *two* photons, one leaving the sheep and one leaving my eye, which meet and annihilate each other at the window. By train time, the two photons appear from nowhere at the window, one travelling inside the train toward my eye and the other travelling outside the train toward the sheep. Neither of these descriptions of how I see the sheep is consistent with what we know of the laws that govern the behaviour of photons.

Yet this does not show that the situation I have just described is logically impossible, since neither of my two descriptions is either self-contradictory or inconsistent with the other one. On the contrary, since the two descriptions are relative to opposite time directions, each of them entails the other. Nor need they conflict with the laws that say how photons behave within regions where time has only one direction. Again, on the contrary, those very laws are what tell us how photons would behave at boundaries between regions with opposite time directions.

We can also avoid these boundary problems by denying that backward time travel entails any process of *travelling*. I said earlier, *if you arrive somewhere, you have travelled there, regardless of how you did it*. In particular, therefore, all that backward time travel, say from Cambridge in 2050 to London in 1950, really requires is that the travellers who leave Cambridge in 2050 must *arrive* in London in 1950: it need not require them to undergo any process of travelling between those two places and times. Admittedly, without the continuity that such a process provides, it may be hard to identify the travellers who arrive in 1950 with those who left in 2050; which is why, other things being equal, I prefer not to let time machines vanish *en route*. But in backward as opposed to forward time travel, other things are *not* equal; it may well prove harder to make sense of the boundary of a backward-travelling *TARDIS* than a discontinuity in its world line. So in case it is, and in order to give backward time travel a run for its money, we should let *TARDIS* vanish *en route* if it has to; as we have noted already, in *Dr Who* programmes it always does.

(It may be worth noting here that a fourth spatial dimension would enable *TARDIS* to avoid causal contact with anything in our three-dimensional space and still keep a continuous world line, just as the two dimensions of the earth's surface let trains from Cambridge to London bypass Hitchin by using the Liverpool Street line. But although higher spatial dimensions have been postulated, we should not rely on them to solve the problems of backward time travel. For if *TARDIS* can move in four spatial dimensions, so can the other things among which *TARDIS* is travelling back in time, a fact that simply regenerates the original problem.)

Future pasts?

Let us agree then that, one way or another, *TARDIS* could travel from 2050 to 1950 without interacting *en route* with the outside world. Even so, other apparent problems remain. First, suppose we ask *when TARDIS* arrives in 1950. It is tempting to say that *TARDIS* cannot arrive in London in 1950 until after it starts travelling in 2050: in other words that, until 2050, the events of 1950 do not include *TARDIS*'s arrival in London, but that after 2050 they do. This indeed is a common feature of tales of time travellers anxious not to affect the past in ways that might threaten their own subsequent existence – as when careful dinosaur hunters only shoot beasts that previous travellers have seen dying naturally an instant later.

The trouble with this, as with all such tales, is of course that it contradicts itself, by implying first that a certain dinosaur died naturally, and then that it was shot; just as our *TARDIS* tale implies first that *TARDIS* did not, and then that it did, arrive in London in 1950. If time travel really entailed such contradictions, it would indeed be impossible. But it does not, since what it implies is not temporal but counterfactual. Dinosaurs need not die naturally 'before' their hunters kill them, and *TARDIS* need not fail to arrive in 1950 until after 2050. All that is meant, and all that backward time travel requires, is that, had the time travel not occurred, the dinosaur *would* have died naturally, *TARDIS would not* have arrived in 1950, and so on. This presents no problem, being quite consistent with the dictum of p. 51, that 'if *TARDIS will* arrive in our past, then it *did* arrive in our past'.

Too many *TARDIS*s?

Another non-problem of time travel is this. Suppose that, as shown in Figure 5, *TARDIS* is made in Cambridge in 2030 but not used until 2050, and then only to go back to London in 2040, after which it remains in London and is never used again. So while, as Figure 5 shows, there is only *one TARDIS* before 2040, and after 2050, in between those two years there are *two*, one in Cambridge and one in London. But how can this be? How can these two different machines be one and the same, as our story implies? Is this not a contradiction, and therefore impossible? If it is, then since backward time travel implies that this *is* possible, it itself must be impossible.

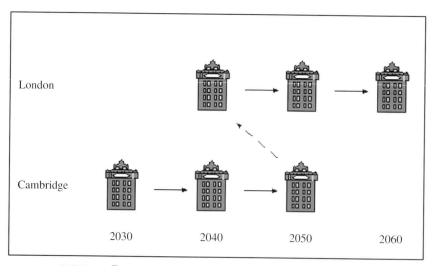

FIGURE 5. Too many *TARDIS*s.

Not so, for unless some other feature of time travel rules it out, I see nothing impossible in Figure 5, merely a counter-example to the thesis that nothing can be in two or more places at once. Normally that is true, as we can see by asking what makes the Cambridge *TARDIS* of 2030 identical with the Cambridge *TARDIS* of 2040. Part of the answer to this question is that the 2040 Cambridge machine depends on the 2030 one for both its features and its existence. For just as we could make the former different – in shape, size, engine, etc. – by making the latter differently, so we could stop the latter existing by not making the former at all. This causal dependence is not of course *enough* to make the 2030 and 2040 Cambridge machines identical (or it would make us identical to our parents!). But it is *necessary*, which is why in general two things A and B that are in different places at the same time cannot be identical – because, since causes must generally *precede* their effects, neither A nor B could then depend on the other in the ways that their identity would require.

But that cannot rule out this case, since causes within *TARDIS do* precede their effects by *TARDIS* time, which after all, as we saw earlier, gets its direction from that of causation inside *TARDIS*. So, while the 2040 London machine is indeed simultaneous with the 2040 Cambridge one, it can still depend on it, in all the ways their identity requires, via its dependence on the 2050 Cambridge machine. And as for whatever else their identity normally requires – for example, a continuous world line linking the Cambridge and London machines – we have already agreed (p. 59) to waive those extra requirements if need be. This being so, the situation shown in Figure 5 raises no new obstacle to backward time travel.

Who made *TARDIS*?

The problem just posed may in any case be avoided altogether, as follows. Imagine a single *TARDIS* vanishing in 2050 to arrive in 1950, when it is put in the Science Museum in London. There it remains for a century, until a group of would-be time travellers wondering how to make a time machine remember that there already is one in the Science Museum, complete with instruction manual. So in 2050 they retrieve it from the museum, read the manual and set off for 1950, where . . . .

Now as in this story, shown in Figure 6, *TARDIS* is never in two places

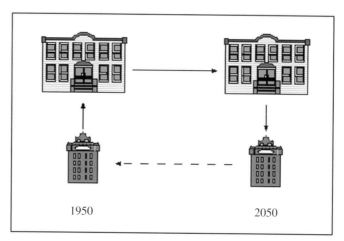

FIGURE 6. When was *TARDIS* made?

at once, the problems, if any, of multiple machines never arise. The problem here is one not of duplication but of origin: when, how and by whom was *TARDIS* made? The answer is of course that it was never made, in any way, by anyone. But is that not an impossible price to pay to stop it ever having to be in two places at once?

Not necessarily: here too I think time travel deserves the benefit of the doubt. It is true that this story, like its predecessor, seems to conflict with what we take to be laws of nature, such as the laws of conservation of mass and energy which *TARDIS*'s appearing in 1950 and disappearing in 2050 would violate. But perhaps these laws, like those governing photons, mentioned above, hold only when and where there is no time travel. And even if, by holding everywhere, they stop time travel occurring anywhere in fact, they themselves may still not be *necessary*; so that time travel *could* occur in worlds with different laws. Here too I see no proof of the absolute impossibility of backward time travel.

Time travel: the insoluble problem

The real objection to backward time travel is at once the best known and the least widely accepted, at least among philosophers. It is the objection hinted at in the dinosaur hunting story, namely that if we could travel into the past we could always cause a contradiction, by doing something then that would stop us setting off now. But no one can cause a contradiction; so backward time travel must be impossible.

The standard reply to this objection is that it proves nothing, because time travellers can also *avoid* causing a contradiction, by acting in ways that *are* consistent with the travel occurring. And certainly the fact that a time travel story *can* be inconsistent does not show that it *must* be. We can after all make any story inconsistent, for example by saying first that Jill and Jane are sisters and then that they are not. If this gratuitous inconsistency does not prove the impossibility of sisters, why should an equally gratuitous inconsistency prove the impossibility of backward time travel?

To see why this reply fails to meet the objection, we must look more closely at what backward time travel entails. To simplify matters, let us imagine not a dinosaur hunt but a time traveller, $Jack_1$, meeting his younger self, $Jack_2$. In itself this duplication of Jacks poses no more problems than the duplication of time machines. But now suppose that $Jack_1$ and $Jack_2$, recognising their identity, fight over which of them should have Jack's property, girlfriend, etc., and that one of them kills the other. If $Jack_2$ kills $Jack_1$ we have no contradiction (merely a long range suicide!). But if $Jack_1$ kills $Jack_2$, that will stop his younger self living to travel back in time and become $Jack_1$, which contradicts the assumption that he does so, i.e. that both $Jack_1$ and $Jack_2$, and hence the fight between them, exist. In short, while the younger Jack could kill the older one, the fight could not possibly go the other way. Yet this fight, if it could happen at all, could surely go either way. Therefore, since it could *not* go either way, it could not occur, and nor therefore could the backward time travel that would enable it to occur.

More precisely, which way this or any other fight can go depends only on its circumstances, including how strong and skilful each fighter is when they fight. Given that a fighter has those attributes when he fights, how he got them, and where he came from (and how), is irrelevant. This is why I said that 'for *TARDIS* and its passengers to be somewhere in 1950, they must be able to interact with the other things and people there then, just as those things and people interact with each other'. And so it is with $Jack_1$. For him to be when and where $Jack_2$ is, he must be able to do to $Jack_2$ whatever anyone who was just like him could do. Thus suppose for example that, in the circumstances, $Jack_2$ would die if shot first by James, who is exactly like $Jack_1$ except that he has *not* arrived from the future. Then $Jack_1$ must also be able to kill $Jack_2$ by shooting him first, just as James can. Yet he could not, since $Jack_2$ cannot be killed by $Jack_1$, whatever $Jack_1$ does, without creating a contradiction. But since nothing can be both possible and impossible – since that is itself a contradiction – this situation could not arise. But it *could* arise if Jack could travel back in time to meet his earlier self. So that too must be impossible.

This is why the fact that we can always tell a consistent tale about what *actually* happens is not enough to prove the possibility of backward time travel. Backward time travel tales must also be consistent with all the facts about what *would* have happened had other things happened, for example that $Jack_2$ would have died if $Jack_1$ had shot him. But they can never be consistent with all such facts, for no travellers from the future can arrive in the past without thereby ruling out some of those facts, namely those whose consequences would prevent the travel occurring. That is what makes time travel into our past impossible and thereby explains, as those who say it *is* possible cannot explain, why it will never happen – as we know it will not, since we know it never has.

FURTHER READING

Earman, J., 'Recent work on time travel', in *Time's Arrows Today: Recent Physical and Philosophical Work on the Direction of Time*, ed. S. F. Savitt, pp. 268–310, Cambridge: Cambridge University Press, 1995.

Grünbaum, A., 'The anisotropy of time', in *Philosophical Problems of Space and Time*, pp. 209–280, London: Routledge & Kegan Paul, 1964.

Harrison, J., 'Dr Who and the philosophers, or time-travel for beginners', *Proceedings of the Aristotelian Society* **45** (1971), 1–24.

Hawking, S. W., 'The arrow of time', in *A Brief History of Time*, pp. 143–153, New York: Bantam, 1988.

Horwich, P., *Asymmetries in Time: Problems in the Philosophy of Science*, Chapters 3, 4, 6 and 7, Cambridge, MA: MIT Press, 1987.

Lewis, D. K., 'The paradoxes of time travel', in *The Philosophy of Time*, ed. R. Le Poidevin and M. MacBeath, pp. 134–146, Oxford: Oxford University Press, 1993 [First published in the *American Philosophical Quarterly*, **13** (1976), 145–152.]

Mellor, D. H., *Real Time II*, Chapters 10–12, London: Routledge, 1998.

Reichenbach, H., *The Direction of Time*, ed. M. Reichenbach, Berkeley: University of California Press, 1956.

4 The Genetics of Time

CHARALAMBOS P. KYRIACOU

Introduction

I expect that, for most readers, the title of this chapter would at first glance seem a little curious. Perhaps it is a misprint, and 'physics' should replace 'genetics'? Most people would correctly consider the scientific study of time to lie more in the domains of theoretical physics and mathematics. These aspects will be covered in other chapters, but here the biology of time, and particularly the question of how time is encoded within the genome of an organism, will take centre stage.

Biological time, be it for a bacterium, a plant, a fruitfly or a human, is represented by any temporally defined activity. For example, circadian time (Latin: *circa* = about, *dies* = day), the major subject of this discussion, is the 24 hour cycle of behaviour and physiology that percolates through the very essence of almost every higher organism that lives on this planet. However, there are many other time scales, both longer and shorter than 24 hours, that have biological significance. Moving up the scale, the ovulation cycles in human females show a monthly rhythm. Our larger domesticated mammals show annual cycles of reproduction that are closely tied to the number of daylight hours, or photoperiod. There are also well-known examples of rhythms that span several years, for example some insect pests show six to seven year swarming cycles. Moving down from the circadian range, there are 60–90 minute cycles in rapid-eye-movement (REM) sleep in humans, 60 second cycles in the love songs of fruitflies, one second cycles of respiration and heartbeat in humans, and millisecond cycles in the way neurons can spontaneously fire in many species. All of these rhythmic patterns are encoded in the genetic potential of the organism, and thus biological time, the fourth dimension, is generated by the three-dimensional DNA molecule.

Ancient history of biological time

The study of biological time can be traced back to the days of Alexander the Great. During his campaigns, the accompanying philosopher Androsthenes noted how plants would raise and lower their leaves towards the sun during the day, then lower them at night. This early observation was confirmed and extended by de Mairan, the 18th-century French philosopher. In controlled experiments, he monitored plant leaf movements in darkened rooms, and found that even in the absence of light, the leaves would still move rhythmically, rising during the subjective day, and falling during subjective night. De Mairan's paper, published in 1729, was the first scientific discussion of biological time, and suggests the presence of an endogenous clock, with a period of about 24 hours, that could maintain the rhythmic leaf movements in the absence of external stimuli. At about the same time, Carolus Linnaeus, the great Swedish botanist and taxonomist, also became aware of rhythmic plant movements. He had noted that different species opened their flowers at different times of day, and his gardens were filled with flowering plants, so he could tell the time simply by observing which of his plants had opened their petals.

Not much happened for 200 years. In the middle part of the 20th century, Erwin Bunning began his classic studies of plant circadian rhythms, Colin Pittendrigh probed the mechanisms of the fruitfly clock by observing its responses to external stimuli such as light and heat, and Jurgen Aschoff placed humans in isolated conditions and monitored their circadian rhythms in physiology and sleep. Circadian cycles were by far the most interesting and pervasive types of biological rhythms, and so researchers tended to focus on these. They were also obviously easier to study experimentally than circannual rhythms for example. Who wants to wait a year to do an experiment?

'Jet-lag' and other problems with clocks

By 1960, circadian rhythm research had almost become respectable, with a famous symposium being held at Cold Spring Harbor in New York during that year. This coincided with the realisation that circadian clocks were important in humans. The term 'jet-lag' had been coined to describe that peculiar malaise that haunts transcontinental air travellers, and seemed to be related to biological clock dysfunction. This feeling of tiredness and disorientation had been first noted by Wiley Post, one of the intrepid band of 1930s American aviators, as he flew the *Winnae Mae* around the world in eight days. Crossing different time zones at speed had only recently been attempted, and its effects on the human circadian clock were, at that time, not appreciated.

In fact, the paradox of time for a traveller on a revolving planet had been described by Antonio Pigafetta, one of Magellan's subordinates, during their epic three year circumnavigation of the globe between 1519 and 1522. Pigafetta kept a diary, recording daily this voyage to hell in which 90% of the crew died, including Magellan himself, who was killed in the Phillipines. When Pigafetta reached the Azores in 1522 after three miserable years, he was surprised to find that the locals had confused the day of the week. They said it was Thursday, when, from his diary, he was sure it was Wednesday. Of course what had happened is that, during his slow movement around the globe, he had passed that arbitrary point, which is necessary on a revolving planet, to distinguish yesterday from today. That point was placed by the good citizens of Greenwich, many years later, at a point furthest from themselves, so as to avoid any inconvenience. Thus the international dateline sits at the other side of Australia.

Of course, Pigafetta's journey was a relatively slow one, and he did not experience any discomfort on crossing time zones as he travelled westwards from Spain (other than that caused by starvation, disease and war). That came 400 years later with Wiley Post. The most dramatic examples of the effects of jet-lag are discovered in the aftermath of air disasters when the preceding work schedules of the air crews are examined. Such was the case of American Airlines flight 808, which crashed in Cuba on the 18 August 1993. In fact in more than 70% of aircrashes, fatigue of the aircrew caused by the desynchronisation between their internal body clock and the external environment is a contributory factor in such accidents. However, jet-lag does not affect only the air traveller. A quarter of the Western working population work shifts, and increasing numbers of reports of ill-health and serious industrial accidents among this group underline the need for clock research to provide some kind of therapeutic interventions. The most tragic industrial disasters – the nuclear reactor failures at Chernobyl and Three

Mile Island and the chemical plant explosion at Bhopal – were caused by shift workers making errors in the early hours of the morning, which corresponds to the trough in the cycle of many human physiological rhythms. The number of car accidents in the USA peaks at this time, when vigilance is at its lowest ebb, as do the number of errors made in any test of performance or information processing.

Findings such as these have stimulated basic research in human circadian physiology and cognitive functioning, particularly from organisations such as the American military, and revealed how circadian cycles modulate almost every aspect of human behaviour. From the medical perspective, notable animal studies include those from the 1970s that showed how anti-cancer drugs were considerably more effective when given at certain times of day. Similarly, the side effects of radiation were significantly diminished when administered at specific phases of the circadian cycle, with, again, obvious implications for human cancer therapy. It is thus not surprising that the big pharmaceutical companies are investing significantly in circadian research.

Basic clock phenomenology

Lest we get too hung up on human circadian behaviour, the one thing that basic research over the years has shown, is that, from bacteria to plants to animals, circadian cycles share so many features that their underlying mechanisms must be very similar. Basic properties of all circadian rhythms are:

1. In constant conditions (usually constant darkness and constant temperature), the rhythm will 'free-run' with a period close to 24 hours.
2. Rhythms that are free-running can be reset by brief light, temperature or even social stimuli. Light stimuli applied in the early subjective evening will usually delay a rhythm, often by several hours, whereas stimuli applied late in the subjective night will advance a cycle. Stimuli applied during the day usually have little effect (the dead zone). The shape of this response to stimuli presented at different phases of the circadian cycle is called the phase response curve (PRC) and is remarkably similar for different types of rhythm in different organisms.
3. Rhythms can be entrained to new temporal environments. Thus travelling from the UK to New York requires a five hour delay to the body clock to fit in with the new light–dark cycle. The time it takes to adjust the clock to the new environment generates 'jet-lag'. However,

eventually we adjust to new time zones. Entrainment of the circadian clock to different temporal situations is closely related to how the clock responds to external stimuli, or the PRC mentioned above.

4. Circadian rhythms are clocks, not thermometers, and the 24 hour period does not lengthen or shorten significantly over a broad temperature range. Imagine a biochemical reaction. Increasing the temperature by 10 degrees Celsius should double the speed of that reaction. The 24 hour periods of circadian clocks do not obey this rule, but are remarkably 'temperature compensated', otherwise they would keep very poor time in a fluctuating temporal environment. This is particularly important for poikilotherms (organisms that cannot regulate their own body temperature).

Genetic analysis of the fly clock – the *period* mutants

Circadian research until the late 1960s tended to focus on how biological clocks responded to various external stimuli, as an indirect method for probing their underlying mechanism. There were a number of competing theories about how such clocks might work at the molecular level, but they were just that, theories. It was not until 1971, when a young graduate student, Ronald Konopka, published his genetic analysis of fruitfly circadian behaviour, that progress began to be made in dissecting the elusive molecular oscillator. Even then, it took almost 20 years for the pieces of the jigsaw to begin to fit together.

Konopka fed flies a chemical mutagen and screened the progeny of large numbers of these mutagenised individuals in an assay to pick out circadian mutants. This assay involved the circadian pupal–adult eclosion rhythm. Briefly, fruitflies go through several stages in their life cycle, from egg to larva to pupa, and finally they emerge from the pupal shell as an adult. When a fly is ready to emerge, it does not do so at any time of day. It waits until dawn, which is when humidity is at its highest. *Drosophila* means 'dew lover', so whoever gave this genus its name was an inspired taxonomist. In Africa, where the fruitfly evolved, it is very important for it to avoid dessication during the middle part of the day, when the sun is at its hottest. The newly emerged fly is particularly vulnerable, because its external cuticle is soft and porous. It needs several hours to tan its exoskeleton, making it relatively impervious to fluid loss. So a newly emerging fly is smart and makes sure that it times its entry into the adult world in the morning. If it is ready to emerge in the afternoon, it waits until the following morning. Thus, if you take a mixed-age population of developing flies, and monitor them in constant darkness and constant temperature, they will show these bursts of emergence every 24 hours or so.

Konopka looked for any unusual emergence patterns in his putative mutants and found three mutants whose circadian behaviour was odd. One showed a 19 hour emergence cycle, another one of 29 hours, and the third was arrhythmic. He then placed individual mutants in a small chamber with food at one end, plus a bung at the other to stop them escaping, and then used an infra-red detector to measure the locomotor activity cycles of the flies in constant darkness and temperature over many days. Under these conditions, flies show a sleep–wake cycle, just as humans do (Figure 1). They walk around during the subjective day, but rest during the night, and recent work has shown that these quiescent periods are, physiologically and behaviourally, very similar to human sleep. These sleep–wake rhythms show a period of about 24 hours even in darkness, reflecting the fly's endogenous circadian clock. Again, the short-period mutant in the pupal–adult emergence assay showed a sleep–wake cycle of 19 hours, the long-period mutant again had a long 29 hour cycle, and finally, the third mutant was an insomniac, and showed little evidence of having any kind of rhythm (Figure 1). In both the population assay of emergence and the individual assay for the fly's sleep–wake cycle, these mutants showed the same change in circadian period, clearly revealing that something significant had been altered in their core clock mechanism.

The three mutations were mapped to the same spot on the X-chromosome, in other words they were alternative alleles of a gene that Konopka called 'period' or 'per'. The normal per DNA had been damaged to generate either the short-period per^s mutant, or the long-period per^L mutant, or the arrhythmic per^o mutant. These results generated a great deal of excitement at the time because it was difficult to believe that one could have a perfectly fit and healthy fly that was nevertheless arrhythmic, as apparently was per^o. In addition, subsequent work showed that the per mutants had similar effects on the timing of a very different type of oscillator.

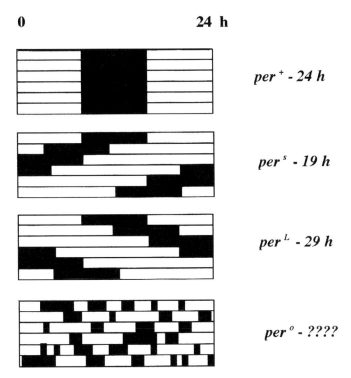

FIGURE 1. Cartoon showing the results of Konopka and Benzer's mutagenesis. Each of the four panels represents a single fly's locomotor activity (black) and rest (white) patterns over six days of continuous recording in constant darkness and constant temperature. The normal fly, per^+ genotype, shows a period of 24 hours in its rest–activity rhythm, beginning and ending at the same time each day. The mutant per^s (short-period) begins its activity several hours earlier on each subsequent day showing a movement of its activity pattern to the left as it cycles every 19 hours. The opposite is shown for per^L (long-period) mutants, whose period is 29 hours. The per^o fly is arrhythmic.

The fly love song – an ultradian oscillator and speciation?

Drosophila love songs are generated by the male fruitfly, as he courts a female, extending his wings, one at a time, and vibrating them to produce an acoustic signal (Figure 2a). This love song carries a train of pulses whose interpulse intervals (IPI) are species specific. Thus a *Drosophila melanogaster* male has a 35 ms IPI on average, whereas the closely related *Drosophila simulans* male sings with a 45 ms IPI. Females were believed to recognise the IPI on their

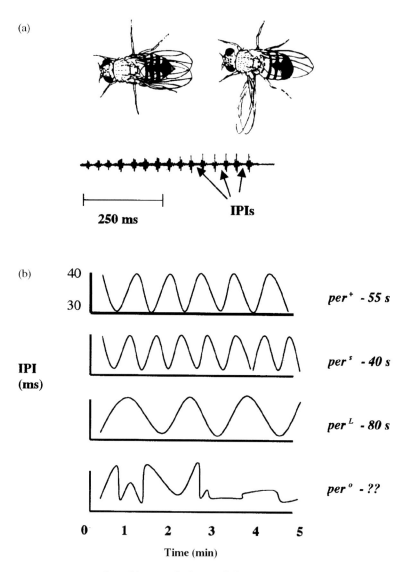

FIGURE 2. Courtship song rhythms and the *per* mutants. (a) A male extending a wing and vibrating it to produce an acoustic signal by which he stimulates the female. A burst of song is shown below the wing which has 15 pulses and 14 interpulse intervals (IPIs). These IPIs average about 30–40 ms in *Drosophila melanogaster*. (b) Over a five minute courtship period, the IPIs show cycles with a period of 55 s in wild-type males. The *per* mutants show love song cycles (or not, see *per^o*) that reflect their circadian behaviour.

conspecific male, and avoid interspecific matings, which would result in sterile progeny, and a loss of their genes in subsequent generations. However, the IPIs are not constant, in that they could begin with longer-than-average IPIs, say 40 ms, then get shorter, down to say 32 ms, then get longer again, then shorter, etc., etc. (see Figure 2b). One complete cycle of IPIs has a duration of about one minute, and then repeats again and again. In *D. simulans*, the closest relative of *D. melanogaster*, oscillations in IPI are also present, but the periodicity is shorter, about 35–40 seconds. Simple genetic crosses between the two species reveals that this species difference in the love song rhythm is caused by differences in genes located on the X-chromosome.

One of these candidate genes is *per*. This is because, remarkably, the per^s male has a shorter 40 second song cycle, the per^L love song has a long 80 second cycle, and the per^0 mutant has no song rhythm at all, and is arrhythmic, just as in its circadian behaviour (Figure 2b). This correlation between the effects of the mutations in the circadian and the ultradian domain shows that the *per* gene encodes a general biological timing component, not limited merely to 24 hour clocks. Perhaps then the *D. simulans per* gene acts as a reservoir for the species-specific love song cycle? Testing this hypothesis requires the *per* genes of the two species to be substituted. This was done after the *per* gene had been cloned molecularly. The *D. simulans* gene was incorporated into *D. melanogaster* hosts whose own *per* gene was inactivated by the per^0 mutation. The *simulans* gene restored the normal sleep–wake cycles of the fly, but, when the song of these genetically engineered flies was analysed, a *D. simulans* rhythm of 40 seconds was revealed, confirming the species-specific role of *per* in this important aspect of mating behaviour, and suggesting a role for *per* in the speciation process. After all, any gene that plays a part in mate choice is likely to be involved in sexual selection, a necessary condition for the evolution of species, as Charles Darwin suggested.

Molecular basis of the fly's circadian rhythms

The *per* gene therefore played an interesting role in this 60 second biological timer. What about its role in the 24 hour circadian clock? The *per* gene was cloned in the mid 1980s, but it took a while for the work to really take off. The first important piece of the puzzle fell into place when the product of the *per* gene, PER protein, was visualised in the fly with the use of anti-PER antibodies. PER was found in a set of neurons in the brain called the lateral neurons, and it seemed to cycle in abundance, with high levels of PER late at night, and low levels during the day. Furthermore, closer inspection revealed that late at night, PER would suddenly move from the cytoplam to the nucleus in these neurons. In addition, the *per* messenger ribonucleic acid (mRNA) was observed to cycle in abundance in the fly's head, with a peak early at night and a trough late at night, so that there was approximately a six hour delay between the peaks of *per* mRNA and protein, so as the PER protein goes up, down comes the RNA, and *vice versa*. This suggested that, as the PER protein moves back into the nucleus late at night, it shuts down its own gene. As the PER protein disappears during the day, it releases its block on its own gene, and *per* transcription begins again. This gave rise to the current model for PER action – the negative feedback loop (Plates I and II).

Proteins that shut down or activate their own or other genes are known as transcription factors and have one thing in common: they contain a region or 'domain' that allows them to bind to DNA. For example, the 'homeodomain' is found in genes that are master regulators of development and can switch on or off whole batteries of other genes. PER, however, does not have any obvious DNA-binding domain, so how might it exert its proposed negative effect on its own gene? It must do it indirectly. PER does have a domain called 'PAS', which is named after PER, ARNT and SIM, the three proteins in which this domain was first identified. PAS is a region that promotes protein–protein interactions, so, for example, in the mammal the ARNT protein sticks to the AHR protein via their respective PAS domains. In the presence of a third attachment molecule, which can be dioxin, the complex can move into the nucleus where it acts as a transcription factor and regulates other genes. This is how dioxin exerts its mutagenic effect on DNA. Therefore, unlike PER, ARNT and AHR have DNA-binding regions called 'bHLH' (basic helix–loop–helix), which allow them to associate with DNA.

If PER has a PAS domain, then it may interact physically with another protein, and perhaps this protein also has a DNA-binding domain that will allow it to bind back to the *per* gene to generate the negative feedback? In fact this turns out to be the case. Mutations in two genes, '*Clock*' and '*cycle*', generate defective locomotor activity rhythms in the fly. When these genes were identified at the molecular level, they were both shown to encode bHLH and PAS domains. In addition, the CLOCK and CYC proteins were shown to associate together via their PAS domain and also attach themselves to a short DNA sequence found just in front of the *per* gene, called an 'E-box' (Plate II). This E-box lies in the *per* gene's promoter, a region of DNA to which proteins bind in order to transcribe the gene into mRNA. When the CLOCK-CYC pair bind to the *per* promoter, *per* mRNA is transcribed, thus making CLOCK-CYC the positive transcription factor pair for the *per* gene. Adding PER protein shuts down *per* transcription because the PER-PAS domain interferes with the CLOCK-CYC PAS domains and prevents them from attaching themselves correctly to the *per* E-box. This simple mechanism provides PER with its negative role on its own transcription (Plate II).

In addition, the PER-PAS domain allows PER to attach itself to another protein called 'TIMELESS' (TIM). TIM protein and mRNA cycle with similar profiles to the *per* gene products, so TIM also appears to negatively regulate its own gene (Plates I and II). As *per* and *tim* mRNA are produced early at night, they move into the cell's cytoplasm, where they are translated into PER and TIM protein. As they reach higher levels late at night, they associate via the PER-PAS domain and move into the nucleus. The *tim* gene, like *per*, also has an E-box, to which the positive factors CLOCK-CYC attach themselves (Plate II). Therefore TIM takes PER into the nucleus, and PER can then also prevent *tim* transcription by disrupting CLOCK and CYC. Not surprisingly, *tim*, like *per* mutants, can alter or obliterate the fly's behavioural rhythms.

TIM is light sensitive. As soon as lights come on, TIM degrades, and a little later, so does PER because PER is unstable in the absence of TIM (Plate II). The implications of this phenomenon are extremely important for understanding how the clock can reset itself to light stimuli. You will remember that earlier I mentioned the PRC (phase response curve). The free-running *Drosophila* circadian rhythm in darkness, like many others, will respond in a very characteristic way to brief light stimuli. A light pulse delivered early at night causes a delay in the rhythm. A light pulse delivered late at night causes an advance. A light pulse delivered early at night will therefore degrade TIM protein, but at this time in the molecular cycle, there is plenty of *tim* mRNA available to regenerate TIM protein levels (Plate I). The time it takes to do this and for TIM levels to return to what they were generates a delay in the molecular cycle. Late at night, the same light pulse will degrade TIM, but as there is now little or no *tim* mRNA available, TIM levels cannot be reconstituted (Plate I). This means that TIM levels are now characteristic of the very low daytime levels, so the molecular oscillation has in effect moved forward in time, generating an advance. This simple and compelling explanation for the PRC based on TIM light sensitivity represents one of the triumphs of molecular biology in explaining a seemingly complex behavioural response.

Related to TIM light sensitivity is another protein called 'CRYPTOCHROME' (CRY). This protein is known to be a blue light receptor in other organisms. When the lights come on, CRY is activated, and it associates with TIM, and disrupts the PER-TIM complex's interaction with CLOCK-CYC (Plate II). In other words, during the day, CRY allows CLOCK-CYC to go back to their job of transcribing the *per* and *tim* genes. CRY therefore derepresses the *per* and *tim* genes during the day. CRY is also believed to be involved in the degradation of TIM in light (Plate II). Lest the reader be puzzled as to how a circadian cycle can free-run in darkness when TIM and CRY are not exposed to light, CRY is probably not required in darkness for the clock to run. In light, it simply enhances the derepression of the *per* and *tim* genes. TIM will also degrade in complete darkness because other circadian-controlled degradation pathways will be activated. Light simply enhances the molecular cycle, which would go on anyway in constant darkness.

Finally, mutations in another gene called '*doubletime*' (*dbt*) also drastically change the period of the fly's circadian sleep–wake cycle. At the molecular level, *dbt* encodes a protein called 'casein kinase' or DBT. As PER protein starts to build up in the cytoplasm, it is phosphorylated by DBT (phosphorylation is what kinases do), and this process, which changes the structure of PER, targets PER for immediate degradation (Plate II). So, early at night, as quickly as PER is produced, it is destroyed by DBT. As TIM levels build up, they block the actions of DBT, and PER can finally accumulate to a level where it can dimerise with TIM and move into the nucleus (Plate II). Therefore DBT generates a critical delay between the peak levels of *per* mRNA and PER protein. Imagine that this delay did not occur – as soon as *per* mRNA was transcribed, it was translated to PER protein, which then moved immediately into the nucleus. The PER protein would feedback and shut down its own transcription very quickly, thus making it impossible to generate a molecular transcriptional/translational circadian cycle. A delay of several hours between *per* mRNA production and the function of the PER protein is required in order to provide the permissive conditions for the feedback loop to have a 24 hour oscillation. Shortening the delay will shorten the period.

The limited number of clock genes we have discussed can therefore generate a molecular cycle within pacemaker cells. The two negative factors, PER and TIM, interact with the positive factors CLOCK and CYC. The delay between *per* mRNA and PER protein required for the oscillation is determined by DBT, while CRY enhances the clock response to light. It appears deceptively simple and undoubtedly other clock components will be identified in the future, for example the proteins that degrade PER and TIM or those that convey circadian information from the pacemaker cells to other parts of the fly. The lateral neuron pacemakers have connections to several regions of the brain. A subset of these neurons also produce a hormone, called 'PDF' (pigment-dispersing factor), which accumulates rhythmically in the terminals of the pacemaker neurons. Eliminating PDF from these neurons with a *pdf* mutation produces behaviourally arrhythmic flies and has the same effect as totally removing them (using another genetic trick). Therefore PDF may be a candidate for the hormonal transmitter that is secreted from these pacemaker cells and transmits circadian information to the periphery of the fly.

This raises the important point that the clock resides not only in the fly's brain but also in peripheral tissues. For example, cycles of PER and TIM can be found in non-neural tissues such as the Malpighian tubules (the fly's kidney), which have circadian cycles of water balance. One clever experiment used the gene for luciferase to investigate exactly where in the body of the fly *per* is expressed. Luciferase is an enzyme responsible for producing luminescence and is found in several organisms, but it is best known for its use in the beautiful night-time courtship displays of the firefly (which is really a beetle). The luciferase gene was placed under the control of the *per* promoter and injected back into fruitflies. The promoter, as discussed earlier, is that part of the gene which controls when and where *per* will be expressed. In the fly, luciferase was therefore expressed in all the tissues in which *per* would normally be expressed, and at the appropriate times. By adding the enzyme substrate luciferin to the fly food, the fruitfly 'became' a firefly and revealed circadian cycles of luminescence. The result was spectacular in that these glowing rhythms were observed in all sorts of tissues, from antennae to wings to legs. Clocks were thus found in a wide variety of peripheral tissues. Even a single Malpighian tubule cell is an autonomous clock.

What about mammalian clocks?

In practice, genetic analysis of mammalian clocks is much harder to do than that of fruitflies. The two organisms that have been used are the hamster and the mouse. A naturally occurring clock mutant, called *'tau'*, was found in the hamster some years ago; it reduced circadian locomotor activity to about 20 hours. Elegant transplantation experiments between mutant and normal hamsters revealed that the suprachiasmatic nucleus (SCN), the part of the brain in the hypothalamus long suspected of acting as the circadian pacemaker, was indeed the anatomical focus for the mutant behaviour. Thus a *tau* SCN that was transplanted into the brain of a wild-type animal (whose own SCN had been removed) generated a short-period cycle in the host. Very recently, the *tau* gene has been identified, and turns out to be a mutant allele of the mammalian equivalent (or homologue) of *dbt*, the kinase involved in the fly clock.

Lest you think this remarkable similarity between the fly and the mammalian clock components is merely a coincidence, a mutant screen in the mouse identified an arrhythmic variant which was named *'Clock'*. When the gene was cloned it turned out to be the mammalian homologue of fly *Clock*. In fact the mouse *Clock* gene was identified before the fly gene, that is why the fly gene is called *Clock*. The mammalian equivalent of CYCLE has also been isolated and is known as 'BMAL1' (or 'MOP3'). There are three *per* genes in the mouse, *mPer1*, *mPer2* and *mPer3*; and two *Cry* genes, *mCry1* and *mCry2*. This duplication of mammalian genes as compared with those of diptera is a common phenomenon that reflects ancient duplications of the mammalian genome. There is also a *tim* gene equivalent in the mouse. By and large, the mammalian clock works in a way similar to that of the fly clock. The mPER components cycle in the SCN and other brain regions, and are involved in the negative regulation of their own mRNA. The major difference is that the *mCry* genes, like *mPer1*, are the main part of the negative loop and repress *mPer* transcription. Thus the mCRYs have lost their photoreceptor function. If you make a mouse doubly mutant for both the *mCry* genes, it is arrhythmic, confirming the role of the *mCrys* in generating the rhythm. On the other hand, a fly *cry* mutant will still show robust circadian behavioural rhythms in constant conditions because CRY acts as a photoreceptor, not as a cardinal clock component, although under some conditions the fly *cry* mutant will show a defective circadian response to light. The different mPER proteins can physically interact with each other and seem to have taken on the role played by fly TIM, so mTIM's function is not clear at present. The mCLOCK and BMAL1 (CYCLE) proteins do exactly what they do in the fly – they are positive transcription factors and bind to E-boxes on the *mPer* genes.

Finally, one of the targets of these clock molecules is the gene for arginine vasopressin, a neuropeptide that is released in a circadian rhythm from SCN neurons. The vasopressin gene has an E-box to which CLOCK and BMAL1 bind. Once bound, the vasopressin gene is transcribed rhythmically. Not surprisingly, in *Clock* mutant mice the vasopressin transcription rhythm is abolished. Adding mPer or mTIM proteins also shuts down vasopressin transcription, revealing that this clock-controlled gene is regulated in exactly the same way as the *mPer* clock genes regulate themselves.

One conclusion of all this work is that if ever one wanted any justification for studying the fruitfly, the conservation of these clock components between the dipteran and the mammal provides an utterly convincing case. In terms of the circadian clock, the mouse (and human) are just bigger flies. However, our knowledge of the molecular basis for the circadian clock is not limited to these higher organisms. Considerable work has been done also on the circadian clock of *Neurospora* (bread mould), which has daily cycles of growth. The key component here is a gene called *'frequency'* (*frq*), which shares many of the features of fly *per* and *tim*, in that the *frq* products cycle and the protein negatively regulates its own mRNA. The *frq* mRNA is also light sensitive, rather like the TIM protein. In Cyanobacteria, which are single-celled organisms that show a cycle of photosynthesis, their metabolism is governed by the circadian clock, and a number of key clock genes have been identified. Again, the negative feedback of these proteins on their mRNA is a significant feature of their expression, and the development of this relatively new unicellular model system will be particularly rapid, given the ease of genetic and molecular analyses of such small organisms.

Evolution of clocks

Given that this book is part of the Darwin series, it seems appropriate to mention something about the evolution of circadian clocks. It is clear that it is the same genes in the fly and the mouse that generate circadian rhythmicity, so the clock mechanism has been largely conserved throughout evolution. This in turn reflects its importance; yet it is interesting that arrhythmic animals, such as the fly mutants *per⁰* or *tim⁰*, are perfectly healthy. This does not mean that once outside the comfortable conditions of the laboratory these flies would survive in the wild. However, it does raise the question of which characteristics of the clock are important for reproductive fitness. We might imagine that, for example, the closer the period is to 24 hours, the less daily resetting would be required, and the less energetically costly this might be. A clock that 'resonates' with the circadian period might therefore contribute to the organism's fitness. This idea, which has been around a long time in chronobiology, was recently tested, directly in Cyanobacteria, and indirectly in fruitflies.

In Cyanobacteria, there are a number of mutants that change the period of the circadian clock. Two of these have periods of 23 and 30 hours, respectively. Cyanobacteria reproduce very quickly, and in a month or so 100 generations or more will have passed. This makes the organism ideal for testing out the long-term fitness effects of changing the endogenous period of the clock. An experiment was performed in which equal numbers of cells carrying one or other of the two mutations were placed in an environment that had a 22 hour day superimposed – 11 hours of darkness and 11 hours of light. At discrete intervals during the following month, the mixed colony was sampled and the frequency of the short- and long-period mutants was assessed (Figure 3). After about a month, the 23 hour mutant had taken over the colony and was present at levels of over 95%. In other words, its reproductive fitness was greater than the long-period mutant in the 22 hour day. This could of course be because the long-period mutant was generally sick and thus the short-period mutant would have outcompeted it in any environment.

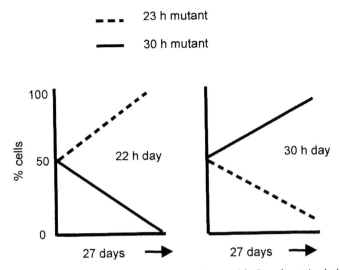

FIGURE 3. Results of the resonance experiment with Cyanobacteria clock mutants. In a 22 hour day environment, the 23 hour mutant (dotted) outcompetes the 30 hour mutant (continuous lines) over a period of a month. The results are reversed in a 30 hour day. The experiment begins with the two mutants in approximately equal proportions.

To test this, the same experiment was repeated, but in a 30 hour daily environment of light–dark cycles. At the end of the month, the frequencies of the two mutants were reversed, with the long-period mutant achieving a frequency of over 90% (Figure 3). Thus the reproductive fitness of the two mutants changes dramatically depending on the environmental light–dark cycle in which they are placed. The take-home message is that, if you are a bacterium and your endogenous circadian period resonates with that of the environment, then the fitter you are in the Darwinian sense.

In the wild, it is unlikely that clock mutants with periods wildly different from 24 hours could survive. However, could natural variation in the period of the cycle have any implications for fitness? In the middle of the *per* gene of *D. melanogaster*, there is a DNA sequence that encodes about 20 pairs of the amino acids threonine (Thr) and glycine (Gly). In the wild, there are two major forms of *per*, one that produces a PER protein with 17 pairs of Thr-Gly, and another that produces a PER protein with 20 pairs. In a protein that has over 1200 amino acid residues, you would think that the extra six residues that distinguish $PER^{(Thr-Gly)_{20}}$ from $PER^{(Thr-Gly)_{17}}$ would be insignificant.

You would be wrong. This natural variation within the *per* gene shows a spatial pattern in Europe. In the south, around the Mediterranean region, very high levels of PER$^{(Thr-Gly)_{17}}$ are found. In the north, Britain, Denmark, etc., the PER$^{(Thr-Gly)_{20}}$ variant predominates. This latitudinal pattern is called a 'cline', and may suggest that natural selection could be favouring each variant in the different environments. However, it could equally be that the cline is simply a result of historical processes. For example, after the last Ice Age, when flies migrated from Africa to Europe, it may have been that the PER$^{(Thr-Gly)_{20}}$ flies were further north to start with, and thus ended further up in northern Europe. However, some esoteric mathematical analyses of the DNA regions around this molecular polymorphism within *per* suggested that natural selection was playing a role in maintaining the two types of *per* gene. If this is the case, then what might be the selective agent?

A first guess would be temperature, because there is far greater annual thermal variation in northern Europe than in the south. In the south it is generally hotter in the summer, and milder in the winter, particularly around the coast. The free-running periods of the two *per* variants were examined under hot and cold conditions. Under hot temperatures, the PER$^{(Thr-Gly)_{17}}$ variant had a period very close to 24 hours, but as the temperature was reduced, its period became significantly shorter. The PER$^{(Thr-Gly)_{20}}$ variant showed a period slightly shorter than 24 hours at hot temperatures, but this did not alter at all when the temperature was reduced. Thus the *per* gene encoding PER$^{(Thr-Gly)_{20}}$ showed the superior temperature compensation. Taken against the background of the cline in Europe, it suggests that, at hotter temperatures, the PER$^{(Thr-Gly)_{17}}$ variant would show better resonance to the 24 hour environmental cycle and thus be at a selective advantage in the south. In the north, however, the PER$^{(Thr-Gly)_{20}}$ variant would show better resonance under the generally colder temperatures and show the advantage. Thus the environment provides the selective agent that may determine the frequencies for the two types of *per* gene in the different areas of Europe. If this is true, then we might predict that in the southern hemisphere the situation should be reversed, with the PER$^{(Thr-Gly)_{20}}$ type predominating in the south, where it is colder. This is in fact exactly what is observed in Australia, with higher levels of PER$^{(Thr-Gly)_{20}}$ in Melbourne as compared with the Great Barrier Reef, confirming the role of the environment in shaping the local frequencies of the different variants for this clock gene.

We would not expect similar selective pressures based on temperature for a warm-blooded animal. In addition, the resonance theory does not explain why some organisms have periods that are considerably different from 24 hours, for example *Neurospora*, whose growth cycles have 21–22 hour oscillations. We can only speculate as to why that might be. Nevertheless, we can see that natural selection will also play a role in shaping the clock to the environment in which it finds itself, and this will vary in its effects from species to species.

Future clock research

The momentum of the molecular approach is irresistible, and progress in identifying clock components in a variety of organisms will continue to be rapid. The clock-controlled genes that produce the circadian output will also be identified. Now that the molecular oscillations of the clock components in the organism can be visualised, for example by using anti-PER antibodies, new insights will be gained into the physiological effects of desynchronisation or 'jet-lag'. Perhaps a jet-lagged fruitfly will show different PER cycles in its brain as compared to its leg, and the time it takes for its peripheral tissues PER cycles to catch up with its brain constitutes jet-lag? Such simple ideas can now be easily tested. There is a great deal of interest from the pharmaceutical industry in developing drugs that might ease the effects of desynchronisation on humans, and clock proteins are obvious targets. If the PER cycle can be instantly reset in humans, maybe that is the key to alleviating shift work problems. The next decade will see whether any of this basic research leads to products that might provide significant advances in health to that proportion of our population whose work rhythms interfere with their biorhythms.

FURTHER READING

Costa, R. and Kyriacou, C. P., 'Functional and evolutionary implications of natural variation in clock genes', *Current Opinion in Neurobiology*, **8** (1998), 659–664.

Konopka, R. J. and Benzer S., 'Clock mutants of *Drosophila melanogaster*', *Proceedings of the National Academy of Sciences of the United States of America*, **68** (1971), 2112–2116.

Dunlap, J. C., 'Molecular bases for circadian clocks', *Cell*, **96** (1999), 271–290.

Lakin-Thomas, P. L., 'Circadian rhythms: new functions for old clock genes', *Trends in Genetics*, **16** (2000), 135–142.

Moore-Ede, M., Sulzman, F. and Fuller, C., *The Clocks that Time Us*, Cambridge, MA: Harvard University Press, 1982.

Ouyang, T., Andersson, C. R., Kondo, T., Golden, S. S. and Johnson, C.H., 'Resonating circadian clocks enhance fitness in Cyanobacteria', *Proceedings of the National Academy of Sciences of the United States of America*, **95** (1998), 8660–8664.

Saunders, D. S., *Insect Clocks*, 2nd edition, Oxford: Pergamon Press, 1982.

Scully, A. L. and Kay, S. A., 'Time flies for *Drosophila*', *Cell*, **100** (2000), 297–300.

Wever, R., *The Circadian System of Man: Experiments Under Temporal Isolation*, Berlin: Springer-Verlag, 1979.

Winfree, A. T., *The Timing of Biological Clocks*, New York: Scientific American Books, Inc., 1987.

Young, M. W., 'The tick-tock of the biological clock', *Scientific American*, March (2000), 46–53.

5 The Timing of Action

ALAN WING

Prediction and timing in motor skills

An impressive skill is that of the waiter who brings a tray of glasses full to the brim to your table and sets the glasses down, one by one, without spilling a drop – either from the one that he is lifting or from the ones that remain on the tray. Since one hand is placing each glass, he has only one hand free to balance the tray. Though it starts evenly loaded, as he takes each glass it necessarily upsets the balance, which requires corrective action if the tray is not to tilt and cause spillage. But note the problem: correction on the basis of sensory feedback takes time due to perceptual and motor delays – say one to two hundred milliseconds to process visual input and a similar further amount of time to select and implement appropriate corrective action – and requires attention. Yet the waiter's attention is probably fixed on determining who is to receive the glass and he does not have time (nor would it look so impressive) to check and correct the disturbance to equilibrium caused by removing the glass. The solution to this problem lies in prediction. If the waiter can anticipate the effect of removing the glass he can implement simultaneous correction.

Predictive adjustment of arm position was demonstrated in an experiment in which each volunteer participant ('subject') was blindfolded, while using one hand to support a weight. The weight was then lifted, either by the subject using the other hand, or by the experimenter, and the resulting change in the position of the hand used to support the weight was recorded. When the subject used the other hand to lift the weight, the supporting hand remained almost stationary. However, when the experimenter lifted the weight, there was a large excursion in position, with the first signs of corrective action only evident after a couple of hundred milliseconds. Evidently, in lifting the weight themselves, subjects implemented a correction simultaneous with their lifting of the weight.

Predictive action requires accurate timing. When the subjects lifted the weight themselves, they could judge the weight being removed and hence estimate the required reduction in the force exerted by the supporting arm. But to achieve a constant position of the supporting hand, the time of reduction must have been synchronised with weight removal and this implies prediction of the time of the action. Of course it might be argued that, if motor commands are transmitted from the brain to the muscles of both arms simultaneously and the delays before the movements occur are equal, then simultaneity of action will 'fall out' directly, with no need for active timing. However, people are obviously capable of predictive timing without such simultaneity, as demonstrated in the study shown in Figure 1.

In this experiment subjects used thumb and index finger of one hand to grasp the vertical sides of a receptacle that received the impact of a dropped ball. The force exerted by the thumb and index finger on the receptacle (the 'grip force') was measured continuously. When the subject's eyes were closed and the impact was unexpected, the grip force started changing some 90 ms after the abrupt increase in vertical load force caused by the impact of the ball. If the subject viewed the drop of the ball, the grip force rose ahead of the impact.

Such anticipatory adjustment, which prevents slip of the receptacle, might use vision to predict the time and amplitude of impact and set the appropriate grip force level based on previous experience. However, even if the subject is blindfolded (preventing use of vision) and drops the ball with the other hand, anticipatory increase in grip force is observed. If drop height is systematically varied, the timing and amplitude of the grip force modulation change correspondingly after a period of learning. This study thus indicates the key role of timing in predictive action.

Variability in explicit timing

In this chapter I review what we know about the control of timing, drawing on research from my own and other laboratories, concerned with the question of how the nervous system organises the timing of movements over intervals in the range up to a few seconds long. This range of intervals is important

The Timing of Action

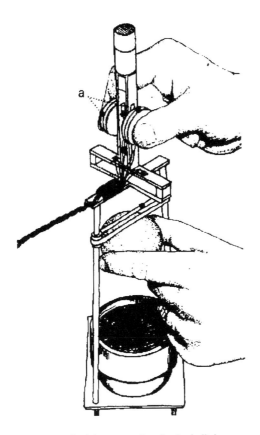

FIGURE 1. Anticipatory action. In the ball drop experiment, a person is asked to hold a receptacle between thumb and index finger (at the points **a** in the diagram). A ball is dropped in the receptacle, and the change in the force exerted by the thumb and index finger is measured. If the drop of the ball is unexpected, the force increases after the ball drops. However, if the person holding the receptacle views the drop of the ball, or if the person is blindfolded but drops the ball himself with his other hand, the force increases ahead of the impact of the ball.

in complex skills, such as musical performance, as well as in everyday actions and so it is appropriate that my starting point is an experiment carried out over 100 years ago by an American psychologist, L. T. Stevens, working at Harvard University. One of the earliest reported scientific investigations into timing of human movement, the work relates to musical performance through its use of a metronome to provide experimental control over the timing task.

In his experiments, Stevens asked the subject to tap a morse telegraph key in time with a metronome, which could be set to produce a beat in the range 60 to 90 pulses a minute (see Figure 2a). Once the subject was following the beat, Stevens stopped the metronome while the subject continued to tap on his own (the so-called 'continuation phase'). Stevens was interested in the accuracy with which subjects reproduced the intervals between the beats in the continuation phase. He was able to measure these intervals with an accuracy of one millisecond. He presented his data in the form of graphs with interval duration on the vertical axis and repetition along the horizontal axis, in fact a time series of time intervals (see Figure 2b).

On the basis of these graphs, Stevens made two fundamental observations. First, he noted that the longer the target interval set by the metronome, the more the length of the time intervals varied in the continuation phase. Second, he observed that this variability had short- and long-term components. He attributed short-term fluctuations around the average to limitations on the accuracy with which motor commands are carried out: 'the hand (or perhaps the will during the interval) cannot be accurately true'. He suggested that the long-term drift around the target reflected 'rhythmic variation of the standard carried in the mind'. In speculating about the causes of fluctuation in interval length, Stevens thus appeared to be distinguishing between lower-level functions related to movement control and higher-level, more psychological aspects of timing.

A number of years ago, working with A. B. Kristofferson at McMaster University in Canada, I proposed a quantitative model of timing, embodying a distinction similar to that proposed by Stevens. We defined a two-stage timing model that contrasted variability due to an adjustable central timekeeper with variability arising in processes responsible for implementing motor commands (see Figure 3a). We hypothesised that the timekeeper is directly controlled by the subject to produce intervals to match the target set by the metronome. In contrast, we assumed that motor implementation involves neuromuscular delays that are not normally under direct control. However, implementation delays may be affected by changing the manner of movement. For instance, tapping the finger using movement of the whole arm might be expected to take longer (and be more variable) than producing a finger tap by flexion of the finger with the arm resting immobile. From a control point of view the Wing–Kristofferson (WK) model is very simple:

The Timing of Action

(a)
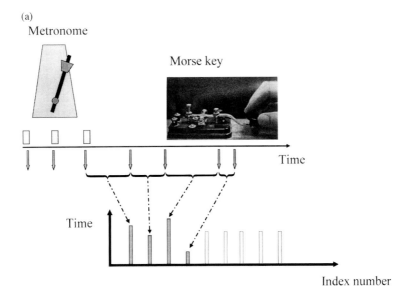

(b)
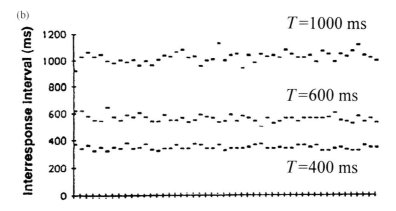

FIGURE 2. (a) Stevens (1886) timing task. A person is asked to tap a morse telegraph key in time with a metronome. The metronome is then stopped, and the person continues to tap on his own, trying to stay as closely as possible to the target interval set by the metronome. The time intervals between the taps are measured and recorded in a diagram, showing the variation in the time intervals produced in the continuation phase. (b) The procedure is repeated with different target intervals and the results are presented in a graph. It is found that the longer the target interval, T, the greater the variation in the time intervals produced in the continuation phase.

it has no feedback loop to correct errors. Indeed, for this reason, it is inadequate as an account of synchronised tapping with the metronome, a point to which I return later. However, it does have the important property that it captures the fluctuating form of the short-term variability noted by Stevens. It suggests that successive intervals between responses will tend to fall on opposite sides of the mean more often than would be expected if the interresponse intervals were varying in a purely random fashion. This property can be explained as follows. Consider a production worker on an assembly line whose job it is to inspect incoming items that are presented regularly on a conveyor belt. Suppose the worker must pick off the items one at a time, inspect each one, label it, and then place it back on the conveyor. Provided inspection times (analogous to implementation delays) are relatively short, the spacing of the incoming items (analogous to timekeeper intervals) determines the spacing of the outgoing items (corresponding to interresponse intervals). Now suppose the worker is slow in inspecting an item (a lengthened implementation delay) – although not so slow as to miss the next incoming item. The slowness will result in an uneven spacing of outgoing items. The space to the slowed item will be longer but, assuming the next item is inspected in the usual time, the following space will be shorter than usual. In a complementary manner, if an item is inspected more rapidly than usual, a short–long pattern of outgoing items results. Thus random fluctuations in inspection time (implementation delay) tend to result in predictable patterning of deviations from the average spacing of outgoing items. Although the inspection fluctuations are random, the process structure results in a degree of predictability in the output intervals. This predictability is captured by the correlation between adjacent intervals. The tendency for short to be followed by long and vice versa results in a negative correlation. In contrast, a positive correlation would have implied that short tends to be followed by short (and long by long) more often than would be expected in an uncorrelated random series.

Many studies have shown that successive time intervals produced in the continuation phase of Stevens-like paradigms are negatively correlated. If one interval is longer than the average (corresponding to the target interval set by the metronome), the following is shorter than the average more often than would be expected for a purely random sequence. This is exactly what is predicted by the WK model illustrated in Figure 3a. In fact, it can be shown for this model that the correlation between successive intervals is bounded between zero (i.e. the lengths of the intervals are no more correlated than the intervals in a random sequence) and $-\frac{1}{2}$ (i.e. strong tendency to alternate between short and long). The actual magnitude of the negative correlation produced by the two-level model depends on the relative amount of the variability of the timekeeper intervals on the one hand and the variability of the motor delays on the other. The larger is the motor implementation variability relative to timekeeper variability, the closer is the correlation to the lower limit of $-\frac{1}{2}$.

Dissociating timer and motor delay

The two-level WK timing model predicts specific relations between the variability of the observed interresponse intervals on the one hand and the variability of the timekeeper and the motor delays on the other. Using these relations, it is possible on the basis of a sequence of observed interresponse intervals to estimate the variability of the timekeeper as well as the variability of the motor delays. To obtain such estimates, an experiment was performed in which subjects produced on each trial a series of responses, with the target interval selected from the range 290 to 540 ms. It was found that the longer the target interval, the higher the variability in the timekeeper, whereas the estimates of the motor delay variability were relatively constant (see Figure 3b). Thus, at longer target intervals, variability reflects the timekeeper, while at shorter target intervals the motor implementation delays are relatively more important (and so the negative correlation between successive intervals is more negative at short intervals).

This situation opens up the possibility of separating the effects of the timekeeper from the processes that implement the motor commands. For example, as the target interval becomes shorter, variability in the implementation delays increases in importance relative to the timekeeper variability. Of course, shortening the interval between two actions eventually results in two simultaneous actions. In this situation, if the motor commands for the two movements are implemented simultaneously, the timekeeper is superfluous (which we suggested might be the case for the initial weight lifting example). Variability in the time of occurrence of one action relative to the other may then be ascribed entirely to the effects of the variability in

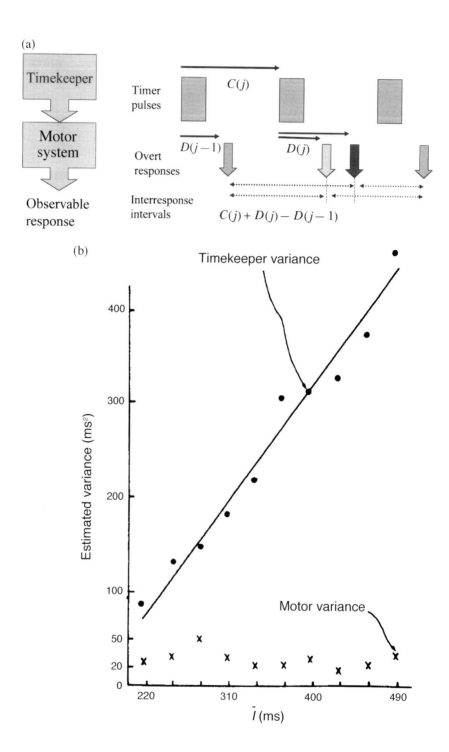

FIGURE 3. (Left) (a) The Wing–Kristofferson (1973) model. According to this model of timing, the observable intervals between the taps arise in two stages: first, a central timekeeper produces intervals to match the target interval set by the metronome. Second, the motor commands that generate timed movements are implemented by the motor system. Thus the timekeeper produces a pulse, which is implemented with a certain delay by the motor system, producing an observable tap. The length of the interval between two taps (indicated by $I(j)$) is determined by three things: first, the length of the interval between two timekeeper pulses (indicated by $C(j)$); second, the length of the motor delay after the first timekeeper pulse (indicated by $D(j-1)$); and third, the length of the motor delay after the second timekeeper pulse (indicated by $D(j)$). (b) The variability in observed interresponse intervals is the result of the variability in the intervals, T, set by the timekeeper, which increases with mean interval, and variability in the implementation of the motor commands, which is relatively unchanged with interval.

the motor implementation processes – any observed variability should be independent of factors affecting the timekeeper.

Such a separation of processes responsible for timekeeping and motor implementation is illustrated by an experiment in which subjects responded to an auditory signal with a pair of responses: pressing a button with their left index finger, followed by pressing a button with their right index finger. With each pair of responses the subject's goal was to produce an interval between responses that matched a previously presented target. Over trials this ranged from 0 to 1000 ms. The experiment was performed under two contrasting conditions: speed and accuracy. In the case of the speed condition the instructions indicated that subjects should respond as fast as possible after the auditory signal, while still aiming to match the previously presented target interval. In the case of the accuracy condition the instructions emphasised that the target interval should be matched as accurately as possible. Under the speed condition, reaction times (the delay from the auditory signal to the button press by the left index finger) were shorter than under the accuracy condition, but the intervals produced were more variable. In fact, in both cases the longer the target interval, the higher the variability of the interresponse interval, but the increase in the variability was slower under the accuracy condition. However, in the case of a zero target interval, demanding simultaneous responses of the subject's left and right index fingers, the variability of the interresponse interval was identical under both conditions. Since in the case of a zero target interval the variability is entirely due to variability in the motor delays, we can infer that the variability in motor delays is unaffected by the two different conditions. On the other hand, the difference in the increase of the variability with increasing target intervals under the two different conditions must be ascribed to differences in the workings of the timekeeper under the two conditions.

Setting the timekeeper

We have already noted that the variability of the timekeeping processes increases with increasing target interval. In Figure 3b the increase is seen to be in direct proportion to the target interval – however, many studies have shown that the variability increases more rapidly than this, in fact, with the square of the mean. But what processes underlie timekeeping? One possible theoretical framework is shown in Figure 4a. It assumes that time intervals are generated by counting rapid neural events until the count matches a target number of events held in memory. If the neural 'ticks' are sufficiently closely spaced, short intervals may be timed by a small target count, while longer intervals are achieved by a proportionately bigger count. Since we are talking about biological timing, it is reasonable to suppose that the interval between neural ticks will vary. Indeed, it has been suggested that the rate of neural events varies with the body's metabolic rate. One experiment even involved trying to speed up or slow down the internal clock, and hence affect timing, by heating or cooling the body! However, even when the mean tick rate is steady, moment-to-moment variability might be expected. If the intertick interval is variable, then the time interval generated in waiting for a given number of ticks will increase in proportion to the mean (Figure 4b).

Variability in the neural event source might not be the only source of variability in timing. Control operations in setting the value of the target count or keeping track of the number of neural events might introduce additional variability. Such control operations are likely to depend on cognitive factors including memory in setting or maintaining the correct target and attention in comparing current and target counts. Certainly timekeeping is a process that depends on the attention paid to a task. This was shown in

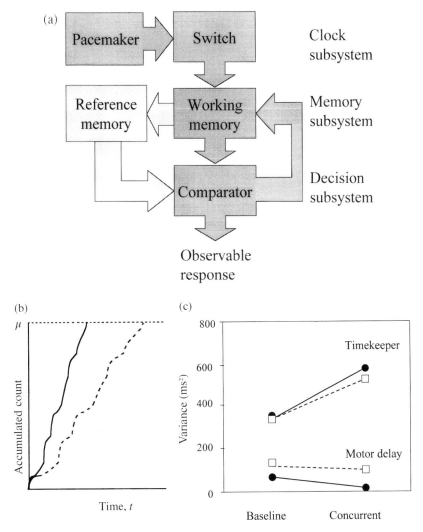

FIGURE 4. (a) Cognitive processes in timing. A pacemaker counter model of timing in which completion of an interval is signalled when the count of pulses matches some stored reference value. (b) The solid and dashed lines indicate how the time taken to attain a given count (μ) may vary from one trial to the next. (c) Taking attention away from timing while tapping a 400 ms interval (by concurrent performance of an anagram-solving task) increases timekeeper variance but not the motor delay variance. Open symbols (dashed lines) indicate estimates from the left hand, closed symbols (solid lines) estimates from the right hand.

an experiment in which subjects were given two simultaneous tasks: they were asked to tap at steady intervals of 400 ms while performing a second task. This second task was a verbal processing task (solving anagrams) and it was found that the attentional demands of performing the second task resulted in more variable timing in the first task (see Figure 4c). Consistent with the two-level model, the increase was due to increase in timekeeper variance, whereas estimates of the variability in motor implementation were unaffected.

So far I have considered timing of actions without detailed consideration of the most commonly used method for achieving experimental control – synchronisation with a pacing stimulus. This very basic skill is fundamental to all kinds of activities, but what does it entail? There are two obvious problems to solve: matching the target interval set by the pacing stimulus, and getting into phase with the stimulus. In the counter model of the timekeeper, matching the interval might be achieved by listening to determine the appropriate event count, and then producing a response each time this number of events is accumulated in the counter. But what about getting into the correct phase? Given delays in motor implementation, the brain must command movement ahead of the expected time of the external pacing stimulus. But how is this lead-time to be set – and note that it will in general vary with the manner of movement and the mechanics of the external system being controlled? According to one possible model for pacing, the perceived asynchrony between each response and the pacing stimulus is used to adjust the timekeeper and hence the phase of the next response. This simple model provides a good account of the form of the experimental data. However, it is instructive to consider some limitations of this model. One is that maintaining good synchrony by applying a strong correction factor implies a cost to performance measured in terms of increased variability of the interresponse intervals. A second limitation is that the correction is taking place within a time period that has already started. This may leave very little time for such adjustment. Indeed it is found that, at higher rates of responding, a better model of performance is provided by adding a second term that reflects synchronisation error sensed over two responses.

Rhythm

The two-level model of timing has been extended to provide a psychological account of rhythm. Western music is frequently organised into rhythmic figures that follow hierarchical rules. Thus bars are subdivided into beats that may be further subdivided into simple fractions of the beat (Figure 5a). This leads to the idea that the production of rhythms may involve separate timers at each level of the hierarchy. If we assume that the variability of the timekeeper at each level is independent of the variability of the timekeepers at all other levels, the model predicts negative correlation, not only for adjacent interresponse intervals, but also for some non-adjacent intervals. Moreover, the variability of the intervals between the repetition of any particular response in successive cycles of the rhythm will be related to the position of that response in the hierarchy. Both predictions received support in an analysis of experiments in which subjects produced synchronous responses with their left and right hands. These experiments were set up in such a way that the motor delays could be ignored, and the covariation (similar to correlation, but without normalisation by the variance) between the left- and right-hand interresponse intervals could be used to estimate multilevel timekeeper properties. However, the analysis of these experiments also led to findings that were not predicted by the basic hierarchical timer model. In particular, in some cases positive rather than negative covariation was found between timekeeper intervals.

One possible account of such positive covariation in interresponse interval is that it reflects propagation of a rate parameter through successive layers of the timer hierarchy. In music, speeding up or slowing down does not affect the fundamental structure of a rhythm. Thus it is reasonable to suppose that rhythms are specified in terms of ratios of intervals (e.g. 1:2) and not by the absolute durations of those elements (e.g. 0.15 with 0.30 s). However, operation of the timekeeper at each level in the hierarchy does ultimately require that the duration be specified. Thus the model assumes that, before each cycle of a rhythm is produced, a preparatory process propagates a rate parameter down through the hierarchy. This involves specifying the interval for the timer at one layer in the hierarchy by multiplying the interval in the next higher level by the appropriate fractional constant to achieve the desired ratio between the higher and lower layer. This mul-

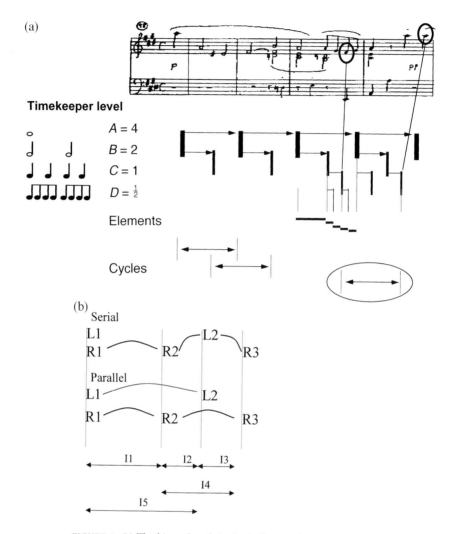

FIGURE 5. (a) The hierarchy of rhythmic figures of Western music. The duration of a bar (in the example given here this is equivalent to four counts) is the highest level of the hierarchy (level A). At the next level down we find intervals equal to half the duration of a bar (this is indicated by level B in the example, equivalent to two counts). At the next level down we have intervals equal to a quarter of the duration of a bar (level C in the example, equivalent to one count), etc. In Vorberg and Hambuch's (1984) extension of the Wing–Kristofferson model, the production of rhythms involves separate timers running in parallel, one at each level of the hierarchy. (b) Interesting rhythms can be produced by generating simple repeated intervals with each of the two

The Timing of Action

> hands, where the intervals are in a non-integer ratio. In the example, the left hand produces intervals of a duration of three counts (indicated by I5), whereas the right hand produces intervals of a duration of two counts (indicated by I1). If a single timekeeper drives the two hands, the target interval for the timekeeper has to be reset continually, whereas if two timekeepers working in parallel are used, each can maintain a constant setting.

tiplicative process introduces overall positive correlations, albeit still modulated by negative correlations reflecting the hierarchy.

The extension of the two-level model to hierarchical production of rhythm involves a set of timers running in parallel. If, instead, the intervals were produced serially by just one timer, that timer would be required to repeatedly switch its setting to the next duration in the rhythm cycle. An interesting but demanding form of rhythm production involves using the two hands to produce simple repeated intervals in each hand, which, because they are in a non-integer ratio (e.g. 2 beats in one hand against 3 in the other), result in complex between-hand patterns of intervals. One characterisation of performing such rhythms suggests that two separate timing systems are at work; the two hands are thus effectively working in parallel. One potential advantage of this arrangement is that two timekeepers, one for each hand, could each maintain a constant setting. If a single timekeeper drives the two hands, the changing intervals between the hands would require continual resetting of the target interval for the timekeeper (Figure 5b). Although the idea of two parallel timing systems seems a natural way to describe such rhythms, it turns out that it is not necessarily correct as a characterisation of how people perform them. With one exception, analyses of the pattern of correlations of the intervals produced within and between hands have shown that polyrhythm performance is sustained by a single timekeeper rather than a separate timekeeper for each hand. The exception, which supports the parallel timing account, is found in the case of skilled pianists performing polyrhythms at high rates.

Analysis of patterns of variation in stable, skilled polyrhythm performance does not generally support parallel timing between the hands. However, it should be noted that models based on separate but interacting timing systems for each hand have been developed that are able to capture certain aspects of the difficulty of producing polyrhythms. Thus, for example, they can account for instabilities that arise in the production of polyrhythms and the way in which one rhythm can spontaneously switch to another rhythm (particularly apparent at higher order ratios such as beating 5 against 6).

Brain function and timing

How is timing of movement controlled by the brain? Studies of the effects of brain damage and functional images of the brain show that movement control involves a distributed network of interconnected brain structures. Primary motor cortex provides the main drive, through neurons coursing down the spinal cord, to the muscles on the opposite (contralateral) side of the body, with different areas of cortex responsible for movements of different parts of the body. A number of other brain regions feed into the motor cortex (see Figure 6). These include the premotor and supplementary motor areas immediately in front of the motor cortex. Lesions in these areas can result in impaired sequencing of action and bimanual coordination, as can lesions in the posterior parietal cortex, which has links to the frontal motor areas. In addition to cortical inputs to the motor cortex there are major pathways via the thalamus to motor cortical areas from the basal ganglia (deep within either cerebral hemisphere) and the cerebellum (at the back and below the cerebral hemispheres). Parkinson's disease, a movement disorder whose symptoms include tremor at rest and slowness of movement ('bradykinesia'), results from basal ganglia dysfunction. Cerebellar damage is often associated with incoordination of multi-joint movement, which, for example, causes difficulty in straight-line movements of the hand.

It will be noted that the schematic of the motor circuits of the brain in Figure 6 contrasts planning and execution functions. It is interesting to ask whether explicit timing skills can be related to specific brain structures in movement planning and whether the two-level WK model can be used to differentiate between those structures contributing to implementation delays and those linked to timekeeping? Two sources of data are available to address this question. The first comprises neuropsychological studies of volunteer subjects with a neurological disorder. In these subjects, patterns of behavioural deficit in timing tasks may be related to the particular areas of the brain that are malfunctioning owing to the disorder. The second comprises brain-imaging

The Timing of Action

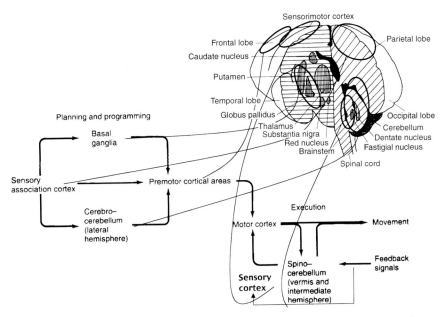

FIGURE 6. Brain and movement. Different cortical–subcortical networks in the brain support planning and execution of movement.

studies in which changes in the pattern of brain activation in healthy volunteers are related to differences between various tasks involving timing.

An example of the first type of data set is provided by studies of the effects of Parkinson's disease. As noted above, the disease symptoms often include not only the well-known and characteristic resting tremor, but also slowness of voluntary movement or bradykinesia. In the early stages of Parkinson's disease, the symptoms are often limited to one side of the body, reflecting a more advanced stage deterioration of function of the basal ganglia on the opposite (contralateral) side of the brain. A neuropsychological study of timing was carried out with a patient in the early stages of Parkinson's disease, when the clinical symptoms were largely unilateral and limited to bradykinesia. The study showed an increase in interresponse interval variability limited to the more bradykinetic side of the body. Analysis of the dependence of the interresponse intervals showed the increased variability was attributable to elevated timekeeper variability, suggesting involvement of the contralateral basal ganglia in timekeeping. A recent group study comparing a group of subjects with Parkinson's disease with a control group of elderly subjects confirmed that increased variability in timing is due to a selective increase in timekeeper variability, with motor delay variability unaffected. While the above studies point to a role of the basal ganglia in timing, other neuropsychological data have suggested involvement of the cerebellum in timing. In a series of case studies of neurological patients with unilateral lesions of the cerebellum, lateralised increases in interresponse interval variability were observed. In this study an interesting contrast was seen between two subgroups of cerebellar patients. In those patients whose lesions affected functioning of the lateral portions of the cerebellum it was shown that the increases in interresponse interval variability were due to raised timekeeper variability. In patients with damage to the medial cerebellum, the increased interresponse interval variability was attributable to elevated motor delay variability, with timekeeper variability relatively unaffected. This result accords with the suggestion in Figure 6 that these two regions of the cerebellum subserve different functions.

Recently, brain-imaging techniques have been used to understand the contributions of different brain regions to timing behaviour. Two studies, using functional magnetic resonance (fMR) images of brain activity have identified cortical and subcortical brain regions contributing to timing. In the first study (see Plate I), 18 second periods of paced responding (target interval set at 300 or 600 ms) followed by unpaced responding with the right hand showed activation in right cerebellum, left sensorimotor cortex and right superior temporal gyrus in both paced and unpaced responding. In unpaced (but not paced) responding, activation of right inferior frontal gyrus with supplementary motor area, putamen (in the basal ganglia) and thalamus was observed. These results led to the suggestion of a fundamental difference between timing in keeping pace with the metronome and continuation timing. It was suggested that the latter involves additional processes to sustain a working memory representation of the target interval, based on right inferior frontal gyrus linked to right superior temporal gyrus, plus explicit timing control based on supplementary motor area, putamen and thalamus.

The production of rhythms with several intervals to be remembered and implemented might be expected to be more demanding on timing and memory than repeated production of a single interval. In a second study using images of brain activity, scans were taken just before reproduction (using the right hand) of a rhythm heard 10 seconds previously. These rhythms included both integer ratio rhythms and non-integer ratio rhythms. The activation pattern in the brain during the reproduction of these two types of rhythm were compared to the activation pattern in the brain during the reproduction of an equal interval series. Significant differences in the activation patterns were observed. Psychophysical tests have shown that non-integer ratios are represented as single intervals whereas integer ratios involve a coherent relational representation. Thus the different activation pattern seen in this second study may reflect greater demands on working memory.

Conclusion

In summary I have reviewed the control of the timing of action. I have shown how fluctuations in the time intervals between a series of movements provides support for a model of timing in which variability due to central timekeeper factors is contrasted with more peripheral factors in implementing each movement. The timekeeper factors relate to more cognitive functions involved in timing, such as memory and attention, while the peripheral factors relate to neuromuscular delays in executing movement. The model is hierarchical in that motor implementation is subordinate to the timekeeper. An extended form of the model, with hierarchical organisation of timekeepers, applies to the generation of musical rhythm. With the addition of feedback processing, the model can be extended to allow movements to be synchronised with an external stimulus. Such adjustments are surely important in the skill evidenced by a musical ensemble playing.

Theories of movement timing have seen recent advances from studies of the underlying brain mechanisms, with data both from neuropsychological studies of brain-damaged subjects and from normal subjects whose brains are scanned as they perform timing tasks. These areas of research are very important and promise to help our understanding of the consequences of brain damage and neurodegenerative diseases. An exciting possibility is that this knowledge may in future help us to better diagnoses and improved therapy programmes for the sensorimotor impairments caused by such diseases, with the prospect of better rehabilitation.

FURTHER READING

Beek, P. J., Peper, C. E. and Daffertshofer, A., 'Timekeepers versus nonlinear oscillators: how the approaches differ', in *Rhythm Perception and Production*, ed. P. Desain and L. Windsor, pp. 9–33, Lisse: Swets & Zeitlinger, 2000. (A comparison of linear timekeeper and non-linear oscillator accounts of timing.)

Hazeltine, E., Helmuth, L. L. and Ivry, R. B., 'Neural mechanisms of timing', *Trends in Cognitive Sciences*, 1 (1997), 163–169. (An interesting introductory overview of issues in timing.)

Kandel, E. R., Schwartz, J. H. and Jessell, T. M. (eds.), *Principles of Neural Sciences*, 6th edition, New York: McGraw-Hill, 2000. (Includes a thorough overview of behavioural and functional anatomical neuroscience of the motor system.)

Vorberg, D. and Wing, A. M., 'Linear and quasi-linear models of human timing behaviour', in *Human Motor Performance*, ed. H. Heuer and S. Keele, pp. 181–262, New York: Academic, 1996. (Detailed technical information on linear models of timing.)

6 Talking about Time

DAVID CRYSTAL

Introduction

When the heroes of Douglas Adams's *The Hitch Hiker's Guide to the Galaxy* arrive at the location described in Part 2, *The Restaurant at the End of the Universe* (pp. 79–80), the narrator pauses for a moment of quiet reflection about the difficulties involved in travelling through time:

> The major problem is quite simply one of grammar, and the main work to consult in this matter is Dr Dan Streetmentioner's *Time Traveller's Handbook of 1001 Tense Formations*. It will tell you for instance how to describe something that was about to happen to you in the past before you avoided it by time-jumping forward two days in order to avoid it. The event will be described differently according to whether you are talking about it from the standpoint of your own natural time, from a time in the further future, or a time in the further past and is further complicated by the possibility of conducting conversations whilst you are actually travelling from one time to another with the intention of becoming your own mother or father.
>
> Most readers get as far as the Future Semi-Conditionally Modified Subinverted Plagal Past Subjunctive Intentional before giving up: and in fact in later editions of the book all the pages beyond this point have been left blank to save on printing costs.
>
> *The Hitch Hiker's Guide to the Galaxy* skips lightly over this tangle of academic abstraction, pausing only to note that the term 'Future Perfect' has been abandoned since it was discovered not to be.

The traditional view

As it happens, '1001' is probably not too far from the truth, at least rhetorically, when we begin to study the ways used by the languages of the world to enable us to talk about time. Certainly we must forget the mindset instilled into most of us when we first encountered the study of the English language in school. There, in a tradition that extends back over 300 years, we would have been told that it is all, really, so very simple. Time is expressed by the verb, through the notion of tense. There are three basic tenses, because there are three logical times along the time-line: past, present, future. Other tenses can make further divisions along this line. We might have 'time completed before the present' – the so-called 'perfect' tense (i.e. the time is 'perfectly past'). We might have 'time not completed before the present' – the so-called 'imperfect tense' (i.e. the time is 'imperfectly past'). Or we might have 'time completed before the past' – the 'pluperfect', a contraction of *plus quam perfectum* (more than the past). Figure 1 illustrates this scheme of things, using one of the earliest and most influential works, Lindley Murray's *English Grammar* of 1795. For anyone brought up on an educational diet of Latin, this would all seem very familiar. In that language, we find exactly the same system, illustrated in Figure 2 for the forms of *amare*, 'to love'. The Latin system looks very neat: each form has a distinctive ending, and indeed it is this concept, of the word-ending changing the time-reference of the verb, that provides the definition for the traditional notion of 'tense'.

But if we compare the Latin and the English systems, with respect to this definition, we encounter a difficulty. Look at the endings in the English examples – or rather, ending, for there is only one. An -ed ending appears in four examples, and the other two have no ending at all: *loved* and *love*. Dare we talk about 'tenses' when there is no distinctive ending? Murray dared, and thus helped to form the orthodox traditional view. He affirms (*ibid.*, Chapter 5, Section 5):

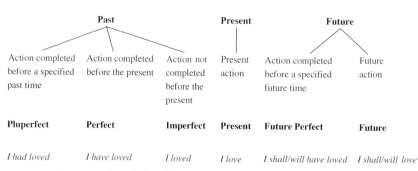

FIGURE 1. A typical traditional analysis of the English system of tenses.

Pluperfect	Perfect	Imperfect	Present	Future perfect	Future
amaveram	*amavi*	*amabam*	*amo*	*amavero*	*amabo*

FIGURE 2. The Latin system of tenses, using the verb *amare* (to love).

> the English language contains the six tenses which we have enumerated. Grammarians who limit the number to two or three, do not reflect that the English verb is mostly composed of principal and auxiliary; and that these several parts constitute one verb. Either the English language has no future tense (a position too absurd to need refutation) or that future tense is composed of the auxiliary and the principal verb. If the latter be true, as it indisputably is, then auxiliary and principal united, constitute a tense.

Other grammarians thought this was a very good idea, and in some works the number of tenses grew and grew, as a result. As the author of the article on 'Grammar', in the first edition of the *Encyclopaedia Britannica* (1771) put it:

> The first and most obvious division of time is into *present, past,* and *future*. But we may go further still in our divisions of time. For as time past and future may be infinitely extended, we may in *universal time past assume many particular times past,* and in *universal time future many particular times future*, some more, some less remote, and corresponding to each under different relations.

And on this basis, he divided each tense into definite and indefinite types – the latter, for example, including *I did write, I may write, I can write* – and arrived thereby at an unspecified but extremely large number of tenses. The influence of this kind of approach is still with us – for example, in the locution 'conditional tense' for *I would write*, which is widely found in the world of teaching English to foreigners.

Of course, there were some who saw through this flagrant disregard of Ockham's razor very early on. Entities were plainly being multiplied well beyond necessity. In 1829 William Cobbett wrote *A Grammar of the English Language* in the form of a series of letters to his 14 year old son James – as the title continues, 'intended for the use of schools and of young persons in general; but more especially for the use of soldiers, sailors, apprentices, and plough-boys. To which are added, six lessons, intended to prevent statesman

[*sic*] from using false grammar, and from writing in an awkward manner.'
The section on 'time' (*ibid.*, §255) begins thus:

> Of *Times* there is very little to be said here. All the fanciful distinctions of *perfect, present, more past*, and *more perfect past*, and numerous others, only tend to bewilder, confuse, and disgust the learner.

And he goes on to explain (*ibid.*, §257):

> Why, then, should we perplex ourselves with a multitude of artificial distinctions, which cannot, by any possibility, be of any use in practice? These distinctions have been introduced from this cause: those who have written English Grammars, have been taught Latin; and, either unable to divest themselves of their Latin rules, or unwilling to treat with simplicity that which, if made somewhat of a mystery, would make them appear more learned than the mass of the people, they have endeavoured to make our simple language turn and twist itself so as to become as complex in its principles as the Latin language is.

A little earlier, he restates the minimalist position (*ibid.*, §255):

> There can be but *three* times, the *present*, the *past*, and the *future*; and, for the expressing of these our language provides us with words and terminations the most suitable that can possibly be conceived.

I can think of several people – not least, from France – who would dispute the last point! But here I want to address only his first observation, that English (and possibly all languages) may be analysed in terms of a simple three-part system. It turns out that the relationship is much more complex – not in the way Lindley Murray or the *Britannica* contributor thought, but sufficiently complex that no-one has yet presented us with a comprehensive account of all possibilities of temporal expression in English, or in any language.

The time reference of tenses

Virtually all traditional grammarians believed that the relationship between time and tense is direct and straightforward. Here is Cobbett again (*Grammar*, §259):

> *Time* is so plain a matter; it must be so well known to us, whether it be the *present*, the *past*, or the *future*, that we mean to express, that we shall hardly say 'we work,' when we are speaking of our *having worked last year.*

The present tense, on this account, refers only to present time – to what is taking place at the moment of speaking. One of the major contributions of the linguistic accounts of English grammar written in the past century has been to demonstrate the fallacy here. It turns out that the present-tense form can refer to virtually any time along the time line. Let us work backwards.

We pick up a newspaper and see the headline, *Jim Smith dies*. It is in the present tense, but the time of the event is recent past: the sentence means 'he has just died'. It does not mean that poor Smith is dying while we read. *He dies*, with present time application, is found only in stage directions. In the newspaper, the use of the present tense gives impact, immediacy. It suggests that the paper is up to the minute. *Jim Smith has died* sounds flat, by comparison. *Jim Smith died* even more so. Figure 3 shows just how many such headline usages there can be, in just part of one page.

Hilary comes in and says to Lucy *I hear you've found a new flat.* But she is not hearing Lucy saying anything. She heard the news some time before – maybe even days before. *I hear you* is possible, as a comment made while someone else is talking, but only in rather special circumstances: for example, there is the acquiescent *I hear you* meaning 'I hear what you're saying (but am not necessarily agreeing with you)', or the confrontational *I hear you, I hear you!*, as an appeal to someone to shut up. The normal use of *I hear* (and *I see*, *I find*, and several others) is to describe the past as if it were hap-

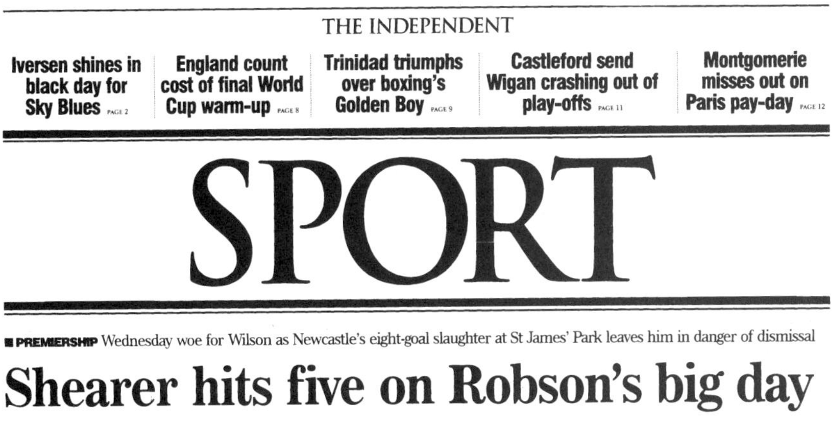
THE INDEPENDENT

Iversen shines in black day for Sky Blues PAGE 2

England count cost of final World Cup warm-up PAGE 8

Trinidad triumphs over boxing's Golden Boy PAGE 9

Castleford send Wigan crashing out of play-offs PAGE 11

Montgomerie misses out on Paris pay-day PAGE 12

SPORT

PREMIERSHIP Wednesday woe for Wilson as Newcastle's eight-goal slaughter at St James' Park leaves him in danger of dismissal

Shearer hits five on Robson's big day

FIGURE 3. A typical newspaper section front page, showing the use of the present tense throughout.

pening now – what is sometimes called the 'historic present'. It is to use the present tense to convey a somewhat dramatic effect.

We can make the present tense refer even further back in time, when telling stories or in imaginative writing. *Listen to this*, begins the raconteur, *I'm walking down Oxford Street, and I see my brother on the other side of the road* .... Only someone who is inexorably perverse, drunk, or under the influence of drugs would respond to this by saying, 'Excuse me, you're not walking down Oxford Street, you're here talking to me'. Once again, for dramatic purposes, the present tense is being used to convey immediacy. The storyteller could have said: *I was walking down Oxford Street and I saw my brother on the other side of the road* ... ; but it distances the speaker from the action. There is no limit to the time gap which can be reported in this way. An imaginative historian might write: *Finally, in 1215, the barons meet the king at Runnymede* .... And it is perfectly normal to see chronological lists written entirely in the present tense: *264 BC First Punic War begins*. Several instances of this can be seen in Figure 4, which is an extract from the 'Chronology' section of *The Cambridge Factfinder*.

What about the present tense being used to refer to future time? *We leave for France tomorrow. I start a new job next week.* Plainly, this is no problem, thanks to the use of adverbial constructions that refer to future time. The adverbial is critical. Nobody will bat an eyelid if I say *I'm leaving for France tomorrow*, and stay sitting in my chair. Not so if I say, *Well, I'm leaving*, without providing any evidence of movement. Eyelids will then be unquestionably batted.

Then there is the use of the present tense to refer to a time reference that extends from the past, through the present, and into the future. This is found when we want to express the idea of an event being repeated, or happening regularly. Here, too, we usually need to rely on an adverbial to express the notion of frequency of occurrence. *I go to town every Thursday* means that on Thursdays in the past I have been to town, on Thursdays in the future I shall go to town ('if I'm spared', as the Irish say), and – if it happens to be Thursday when I am making the remark – I might be going to town while I speak. Sometimes, the sentence has a 'habitual' meaning without any adverbial expansion. *James drinks.*

This last usage is very close to the 'general truth' meaning of the present tense. Here the time frame is extended to include all conceivable times – in

1916 Battle of the Somme.
1916 Irish rebellion (to 1921).
1917 US Expeditionary Force in Europe.
1917 Russian Revolution.
1917 Civil war in Russia (to 1922).
1917 Balfour Declaration promises Jews a home in Palestine.
1918 Fourteen Points statement by President Wilson.
1918 End of First World War.
1918 Women over 30 given right to vote in Britain.
1919 May 4th movement in China.
1919 Foundation of Soviet Republic.
1919 Amritsar massacre in India.
1919 Bauhaus movement established in Germany.
1919 John Alcock and Arthur Brown make first Atlantic air crossing.
1919 First woman MP in House of Commons (Lady Astor).
1919 Adolf Hitler founds National Socialist German Workers' Party.
1919 Spartacist rising in Berlin crushed.
1919 League of Nations established.
1920 Radio broadcasting begins.
1921 Treaty partitions Ireland.
1922 USSR established.
1922 Benito Mussolini in power in Italy.
1922 Frederick Banting and Charles Best isolate insulin.
1922 BBC makes first regular broadcasts.
1922 Tomb of Tutankhamun discovered in Egypt.
1923 Munich putsch by Adolf Hitler.
1923 Republic proclaimed in Turkey.
1923 Major earthquake in Japan.
1924 Death of Vladimir Ilyich Lenin.
1925 Publication of Adolf Hitler's *Mein Kampf*.
1926 General Strike in Britain.
1926 Jiang Jieshi (Chiang Kai-shek) leads movement for reunification of China.
1926 John Logie Baird demonstrates television.
1927 Talking pictures begin.
1927 Charles Lindbergh's first solo flight across Atlantic.
1927 Duke Ellington begins playing at the Cotton Club.
1928 Alexander Fleming discovers penicillin.
1928 Walt Disney introduces Mickey Mouse.
1929 Wall Street crash.
1929 Lateran Treaty establishes Vatican as state.
1930 Amy Johnson's solo flight, England to Australia.
1931 Creation of republic in Spain.
1931 Japanese occupy Manchuria.
1931 Empire State Building built in New York.
1932 Foundation of Kingdom of Saudi Arabia.
1932 Chaco War between Paraguay and Bolivia (to 1935).

1933 Franklin Roosevelt introduces New Deal.
1933 Adolf Hitler becomes Chancellor of Germany.
1933 Reichstag Fire in Berlin.
1933 Prohibition repealed in the USA.
1933 Discovery of polythene.
1934 Long March of Chinese Communists begins (to 1935).
1934 Discovery of nuclear fission.
1935 Italian invasion of Abyssinia (Ethiopia).
1936 Beginning of Spanish Civil War (to 1939).
1936 Anti-Comintern Pact between Japan and Germany.
1936 Arab revolt in Palestine.
1936 British constitutional crisis over Edward VIII.
1936 John Maynard Keynes publishes his economic theory.
1936 First public television transmissions in Britain.
1936 *Queen Mary*'s maiden voyage.
1936 Crystal Palace destroyed by fire.
1937 War between Japan and China begins.
1937 Pablo Picasso paints 'Guernica'.
1937 Golden Gate Bridge completed in San Francisco.
1937 *Hindenburg* zeppelin destroyed by fire in USA.
1937 Jet engine tested.
1938 Germany occupies Austria.
1938 Munich Agreement.
1938 Discovery of nylon.
1938 Chester Carlson makes first xerographic print.
1939 Germany invades Czechoslovakia and Poland.
1939 Second World War begins.
1940 Evacuation of Dunkirk.
1940 Battle of Britain.
1940 Plutonium obtained by bombardment of uranium.
1941 Germany invades Russia.
1941 Japanese attack Pearl Harbor.
1941 Death of James Joyce.
1941 Orson Welles makes *Citizen Kane*.
1942 Construction of first nuclear reactor.
1942 Defeat of Germany at El Alamein.
1942 American defeat of Japan at Midway.
1942 Anglo-American landings in North Africa.
1943 Surrender of German army at Stalingrad.
1943 Capitulation of Italy.
1944 D-Day landing in Normandy.
1944 Education Act in Britain.
1945 Atom bombs dropped on Japan.
1945 Second World War ends.
1945 Yalta Conference.
1945 Nuremberg War Crimes Tribunal opens.
1945 United Nations established.
1945 Republic of Yugoslavia established under Tito.
1946 Perón in power in Argentina.
1946 Civil War in China (to 1949).
1946 Civil War in Indo-China (to 1954).

FIGURE 4. An extract from the 'Chronology' section of the *Cambridge Factfinder*, illustrating the widespread use of the present tense.

other words, the statement is timeless. *Oil floats on water. Two and two make four*. Here, the use of some of the other tense forms is anomalous. *Oil floated on water last week*? *Two and two will make four tomorrow*?

Quite plainly, there is no straightforward correlation between the use of a present-tense form and the reference to present time. One linguistic form can have several time references.

The expression of future time

The opposite also holds: one time reference can be expressed by several linguistic forms. Future time is an excellent domain from which to illustrate this point, because in English it is not tied to a single ending. Strictly speaking, if by 'tense' we mean a system of verb endings chiefly expressing time, then there is no future tense in English, unlike in French (*Je donnerai*), and many other languages. Lindley Murray, you will recall, thought the view that there is no future tense in English to be 'a position too absurd to need refutation'. For him, *I will/shall go* counted as a future tense form. But there are serious problems with this view.

The main problem is this. If we allow *will* and *shall* to be counted as a future tense, because they express an element of future meaning, then we must logically include all the other forms in the language that also express an element of future meaning. There are many of these. Here are the chief contenders.

Alongside *will* and *shall* we have *would* and *should*. *If I went to Paris, I would go up the Eiffel Tower* may be hypothetical, but it is undeniably future. *He should be arriving by boat* likewise. If we insist on calling words and constructions that express future time 'tenses', then this would have to be called a 'hypothetical future tense', or some such.

The very common informal usage *be going to*, as in *I'm going to get something to eat*, typically pronounced /gonə/. This construction allows us to express the notion that an event will take place very soon. On similar grounds, this would have to be called a 'near-future' tense.

The rather less common *be about to*, as in *I'm about to get something to eat*. This construction allows us to express the notion that an event will take place even sooner than *be going to*. It would, I suppose, have to be called an 'even-nearer-future' tense.

The rather more formal *be to*, as in *I'm to get something to eat* – in other words, someone has given me this instruction. Again, it is near future, though whether it is nearer than *be going to* or not is debatable. The time reference is probably very similar; it is the attitude involved that is different. A tense merchant might worry greatly about this, and try to persuade us that one points to the time-line in a different place from the other.

There are many other verb forms in English that express an attitude along with an element of future time. Think of *may* and *might*. *I may go, I might go*. These are plainly future, though the dominant notion they express is possibility or permission. A 'putative' future tense, perhaps? Or two, really, because *may* is not the same as *might*, referring to a greater likelihood of something happening. A 'definite putative' tense versus an 'indefinite putative' tense, doubtless.

We are up to six 'future tenses' now, alongside *will/shall*, but we have by no means completed all the possibilities, even in the verbs (e.g. *have to, had better, have got to*). And we have yet to consider all the adverbial expressions that are capable of expressing future time. Several are future time only: some refer to the very immediate future (*any moment now*); some refer to various kinds of removed times (*in a few minutes, later this afternoon, tomorrow, the day after tomorrow, next week, the week after next, next month, next year*); and some express varying levels of definiteness (*in 27 minutes time, next Monday* versus *one of these days, in due course*). It would be ridiculous to try to turn all of these into tenses.

The *reductio ad absurdum* of this approach is when we find the other tense forms, present and past, being used to express the future. We have already seen how the present tense can be involved in the expression of future time, when an appropriate adverbial is present (*We leave for France tomorrow*). But even the past tense can be used in this way. Consider: *I was going to Paris next Tuesday, but I'm not now*. Past tense, *was* going, referring to next Tuesday, but the event *not* happening now – a non-future future, in effect.

There is a second *reductio*, the complementary of the point just made. Not only do other tense forms express future time, but the two supposed future tense forms, *will* and *shall*, themselves express times other than the future. What does this next sentence mean? *John **will** keep coming in at midnight*. The intonation leads you into an interpretation of past time. This sentence can only mean that John has been routinely coming in at midnight in the past (and it is likely that this pattern of behaviour will continue). It does not mean that John is going to start coming in at midnight at some future point. Or consider this sentence: *Oil will float on water*. Here too, the sentence does not mean that, at some future point, oil is going to start floating on water. Once again, we have a 'timeless' expression.

The problem is evident. If tense is simply a matter of expressing time, then we have to recognise dozens of tenses in English. And the same reasoning would affect other languages. So it cannot be that way. It would make a nonsense of the useful notion of tense. The verb cannot take so much weight. Plainly what is happening is that other bits of the language – auxiliary verbs, semi-auxiliary verbs, adverbs, adverbial phrases – are contributing to the expression of time. Putting this another way: the linguistic expression of time spreads itself throughout the whole of a sentence. Some sentences illustrate this very powerfully. Just reflect on the temporal nuances that you find here:

> The former president is determined to keep on popping in and out of his brand-new office on Thursdays for the foreseeable future.

Past time in *the **former** president*. Present time in *is determined*. Ongoing (habitual time) in *keep on*. *Popping* – a momentary verb, which implies a very short period of time. *In and out* – a frequentative expression, which implies a longer period of time. ***Brand-new*** *office* – with a time-restricted adjective. *On Thursdays* – another frequentative expression. And *for the forseeable future* – a future time adverbial phrase. The time of the sentence moves from past through present to future. 'Time is so plain a matter'?

We are so used to thinking of verbs as the tools a language uses for the expression of time that we forget about the other parts of speech. I have already mentioned adverbs. Here are some examples of the others.

> Adjectives – *brand-new, old, fledgling, mint [condition], experimental, modern, latter-day, up-to-date, topical, traditional, ancient, bygone, obsolete, elapsed, brief, outgoing, punctual, eventual, venerable* – and of course the words *past, present* and *future* themselves.
>
> Nouns – *date, hour, millennium, epoch, morning, day, week, year, season,* etc. Also their proper names – *January, Thursday,* etc., as well as general notions such as *tenure, period, interim, lull, interlude, adjournment, perpetuity, delay, aftermath, successor, occasion, relic, fossil.*
>
> Prepositions – *during, throughout, until, up to, before, after, since.*
>
> Conjunctions – *when, whenever, while,* as well as many items that can also function as prepositions, such as *until, before, after, since.*

> Even parts of words – affixes – can express a time relationship. In English this is chiefly done through prefixes, as in *ante-, proto-, pre-, post-, ex-, fore-, re- (rebuild), neo-, palaeo-*. We might stretch the notion to include the causative suffixes, such as *-en* (as in *frighten*) and *-ify* (as in *beautify*).

We have to conclude that there is no obligatory association between time and the verb.

Other languages and cultures

So far we have discussed time with reference only to English. We must not of course assume that other people, speaking other languages, will think of the time-line in the same way that we do, or even think of time as a line at all. The Amerindian language Hopi has three tenses, but they are not past, present, future. One tense is used for expressing general truths (such as 'Rivers flow fast'), one is used for reports of known or very likely happenings ('I saw her last week', 'I can see you now', 'I shall be with you in a minute' all use this form), and one is used for events that are uncertain ('She is arriving tomorrow', 'They will catch a moose'). These notions cross-cut our own concepts of time, and interact with other grammatical notions, such as aspect and mood.

The ending that might be best associated with tense may appear on parts of speech other than the verb. In Potawatomi, a noun may take an ending that places it in past time: 'my father' versus 'my deceased father', 'my canoe' versus 'my stolen canoe'. This is not the way we express our sense of time, but it is just as logical. We attach time categories to actions, through the verb. There is no reason why other languages should not attach them to things, through the noun. If we want to express the thought that my father has died – in other words father + exist + PAST – we can do it either by attaching the pastness to the action (died) or to the entity affected (as it were, the no-longer-existing-father). English can do this only in fun: do you remember Monty Python's, 'It's an ex-parrot'? In some languages it is the normal way to talk.

In Japanese, time relations can be found on the adjective as well as the verb. In this language, in an analysis made by the American linguist Bernard Bloch, adjectives are inflected for nine categories. One of these (the usual one cited in dictionaries) expresses non-past time: the attribute (e.g. 'good') is true now or in the future. The adjective does not mean merely 'good', but 'BE good' – that is, 'is good now' or 'will be good'. Another inflection expresses past time – the attribute 'was/has been/had been (etc.) good'. Another contrast distinguishes an indicative meaning from a presumptive one: for present time, the attribute is 'probably good, will probably be good, may be good', etc.; for past time, it is 'was probably good', 'must have been good', etc. And so on – nine contrasts in all. Several of the endings correspond to those used in verbs, making the adjective a much more 'active' part of Japanese speech than it is in English.

These linguistic differences relate to the formal ways in which languages express time relationships. They are central to the domain of grammar – its morphology and syntax. And, according to some, their significance goes well beyond grammar. For example, George Steiner, in *After Babel* (1975, p. 132), considers tense forms to exercise considerable control over our whole mindset:

> the inflection of verbs as we practise it has become our skin and natural topography. From it we construe our personal and cultural past, the immensely detailed but wholly impalpable landscape 'behind us'. Our conjugations of verb tenses have a literal and physical force, a pointer backward and forward along a plane which the speaker intersects as would a vertical, momentarily at rest yet conceiving itself as in constant forward motion.

This characterisation betrays its cultural origins, in its Newtonian metaphor of time as a line along which we progress, which we segment into durational quantities, and which we use to schedule things. But the general point is instructive, that the way we talk about time tells us something profound about how people think and how they live. So let us look at some alternatives.

Not everyone talks (thinks) of time in terms that can be related to a single dimension – a line, or path, or road. North American Indian people – the Hopi and Blackfoot, to take two reported examples – do not do so, nor do several peoples in Africa. For them, time is animate, alive, the activity of spirit. Time is what happens when things change. Among the Tiv of Nigeria, for example, time (according to the anthropologist Paul Bohannon) is like a capsule. There is a time for cooking, a time for visiting, a time for working, and, when people are involved in one of these times, they do not shift to another. The day of the week, for instance, is named after the things that are being sold in the nearest market – as it were, Monday is furniture day,

Tuesday is cattle day. This means that, as you travel around, the names of the days of the week keep changing, depending on where you are. Cattle day might be (in Western terms) Tuesday in one part of the country but Thursday in another. You take two days to travel 80 kilometres, and find yourself linguistically on the same day as when you started out.

English does not routinely talk about time in terms of the way things change in the real world. We do admittedly sometimes encounter it through translation from other cultures: 'To everything there is a season, and a time to every purpose under heaven' says Ecclesiastes, and the writer goes on to list various options. But apart from this, the nearest we get to it is in the use of certain idioms, some of which seem to reflect a rural existence in which local events played a critical part. There is a hint of the Tiv way of life when we say *You can keep on saying that till the cows come home*, but it is only a hint. Just a few other time idioms show the possibilities: *till I'm blue in the face, at the drop of a hat, on the spur of the moment*. But they are marginal to the system of English expression.

Another kind of difference concerns the precision and explicitness with which many peoples talk about time. Because things do not change in exactly the same way, because circumstances alter, a language may not express a time system as an exactly repeating cycle of points (60 seconds, 60 minutes, 24 hours, 7 days, 12 months, 10 years, 100 years), and notions that are dependent upon this system, such as appointments, agreed starting times, and the like, do not make sense. This is far removed from our way of talking about time. Precision and explicitness are the bases of our mindset. It would be inconceivable for us to agree to meet without saying when, or to arrange a meeting without saying when it will start. However, such inexplicitness is common in many parts of the world. Edward T. Hall reports several such cases in *The Silent Language*. Here is one, from Afghanistan (*ibid.*, p. 29):

> A few years ago in Kabul a man appeared, looking for his brother. He asked all the merchants of the market place if they had seen his brother and told them where he was staying in case his brother arrived and wanted to find him. The next year he was back and repeated the performance. By this time one of the members of the American embassy had heard about his inquiries and asked if he had found his brother. The man answered that he and his brother had agreed to meet in Kabul, but neither of them had said what year.

A second example. We would think it insulting or incompetent to schedule two or more meetings at the same time. If I were to say to you, 'I'll meet you to discuss your paper at 2.30 p.m. tomorrow' and then say to someone else in your earshot 'I'll meet you to talk about the finances at 2.30 p.m. tomorrow', you would feel affronted. You would say, 'But you're already meeting me at that time'. Anyone who persistently broke the 'one meeting at a time' rule would be considered inefficient. But this is not so in many parts of the world. In some parts of Latin America, for example, it is common to find that several other things are going on at the same time. Edward Hall again (*ibid.*, pp. 19–20):

> An old friend of mine of Spanish cultural heritage used to run his business according to the 'Latino' system. This meant that up to fifteen people were in his office at one time. Business which might have been finished in a quarter of an hour sometimes took a whole day.... The American concept of the discreteness of time and the necessity for scheduling was at variance with this amiable and seemingly confusing Latin system. However, if my friend had adhered to the American system he would have destroyed a vital part of his prosperity. People who came to do business with him also came to find out things and to visit each other. The ten to fifteen Spanish-Americans and Indians who used to sit around the office ... played their own part in a particular type of communications network.

It is not the English linguistic way to dispense with temporal precision. The nearest we get to it is in colloquial speech, when we say such things as *it'll take years* (where we do not mean literally years) or *we'll be home in a while*, where *while* means different things to different people. The language does allow us a certain amount of fuzziness – *I've been here for ages, for yonks, I'll do it in due course, one of these days* – but they are marginal to the system.

We have to recognise that cultural differences enter deep within the system of time expression in a language. As Guy Bellamy puts it, in *The Comedy Hotel* (1992, Chapter 12):

> A French five minutes is ten minutes shorter than a Spanish five minutes, but slightly longer than an English five minutes which is usually ten minutes.

And there are even more radical cases, where conceptions of past, present and future interact in more profound ways. The Aboriginal dreamtime is from the remote past, but is still alive, present, and accessible to modern members. Indeed, ways of talking about time in Australian Aboriginal languages present many differences from those familiar to Westerners. Some languages have the same word for 'today' and 'tomorrow', for example Ngiyampaa has *kampirra* meaning 'a day on either side of the reference time'. Eastern Arrernte has the same word, *apmwerrke*, for 'yesterday, a few days ago, in the last few days'. Wik-Mungkan uses *peetan* similarly.

Time expressions we live by

People sometimes express surprise that a language might not distinguish between yesterday and tomorrow. Such cultures can't have a very well developed sense of time, they say. This evolutionary way of thinking is misconceived. We must not fall into a Whorfian time-trap (see *The Cambridge Encyclopedia of Language*, 1997, Chapter 15): it is not that these people have no sense of the passing of time, or that time is not of importance to them. It is simply that their language encodes those aspects of time that they find to be of importance in carrying on their lives. If the distinction between yesterday and the day before is not of importance, it does not need separate words or inflectional endings. One might be able to ingeniously express the difference, but it is not routine.

We are the same. For us, 'yesterday', 'today' and 'tomorrow' are important, so we distinguish them, and we have standard locutions for 'the day after tomorrow' and 'the day before yesterday'. But we do not have a series of separate words for, say, different parts of the month: for us, the first week of the month, the second week of the month, and so on, are not usually important, especially in a system where weeks do not divide neatly into months, and where the months have limits that are so arbitrary we have to teach ourselves rhymes in order to remember how many days they have ('Thirty days hath September...'). Cultures that carry out activities on a strictly lunar monthly basis are likely to have developed an appropriate language to talk about it. Indeed, if a culture finds any particular time critical, then its language will reflect it in its lexicon or grammar. The Australian Aboriginal language Meryam Mir has the expression *koki kerkerge*, meaning 'in the middle of the north-west wind time' (i.e. the monsoon season). Such expressions would be of little value in the UK: what would we do with a conventional expression for 'in the north-wind season'?

You can always tell which temporal domains are not viewed as central by a culture through the absence of expressions for talking about that domain, or through the way words grow in imprecision and ambiguity as they approach that domain. Evidently times further backward and forward beyond two days are not so important to us, for we have no standard locutions for them, and ambiguity can emerge. Weekly reference is fuzzy, for example. When exactly is 'a week ago'. I am speaking to you now, on Friday. I say that something happened a week ago. Must it mean 'last Friday'? Did an event last Thursday not also happen a week ago? How far back may I go before it becomes 'two weeks ago'? And if it happened last Saturday, which is less than a week, was this not also a week ago? Or again, I just said 'last Thursday', and you took me to mean a week ago. But strictly, in this case last Thursday was yesterday. To be precise, I might have added 'a week yesterday'. The same problem applies to 'next'. On Monday I say 'I shall see you next Friday', and we have no problem. 'I shall see you next Tuesday' must, however, mean 'a week tomorrow'. So when is the boundary line? It is unclear. 'I shall see you next Wednesday' is ambiguous, and the exact date of the appointment had better be checked.

By contrast, the terms and locutions that are frequently used in a language do tell us something about the mindset of a culture. It is illuminating to examine the temporal metaphors we live by. Which verbs typically accompany the noun *time* in English? They are metaphors of value and ownership: we *have* time, *find* time, *allow* time (for something), *take* time, *give* time, *fix* time, and *borrow* time (by *living on borrowed time*). We *need* it, *spend* it, *save* it, *waste* it, *lose* it, *gain* it, *buy* it, *value* it, *make* it *up*, and *play for* it. There are metaphors of speed and measurement: time *passes, whiles away, flies, runs, drags, hangs (heavily)*, or *stands still*; we can *mark* time and *keep* time. There are a few metaphors of creation and death. We can *make* time. Time can *heal*. ('Time wounds all heels', as Groucho Marx said.) And, if we don't like time, we can *kill* it ('before it kills us', as Herbert Spencer once added).

It does not have to be this way. All kinds of other metaphors could be used to talk about time. Ludic metaphors, for instance. We might play with time, or sport with it (as does Sanskrit). We might construct with time – building or demolishing (as in some South Slavic languages). We might give it an aesthetic expression: time might sparkle or look nice, be clean or unclean. Time might have physical or biological properties (wet or dry, male or female).

It might be sensual – auditory (listen to its sound), visual (see its colour or shape), tactile (feel its quality), olfactory (smell it), gustatory (taste it), telepathic (sense it). We lack a comparative idiomatology of time expressions. Unfortunately, we lack a comparative idiomatology of anything.

The literary dimension

But, you might be thinking: are there no auditory verbs used in relation to time? What about this sequence from the First Voice's opening monologue in Dylan Thomas's *Under Milk Wood*?

> Time passes. Listen. Time passes.

Here an auditory verb is being made to collocate with *time*. That, in a phrase, is what creative authors are for. They are there to break the rules. But the rules have to be there first. As Robert Graves once said, 'A poet has to master the rules of English grammar before he attempts to bend or break them.' People like to play with time expressions, and when we examine a literary corpus we begin to see the way in which people can break out of their Western mindset and make contacts with those of other cultures. T. S. Eliot reaches out towards an Aboriginal conception of time when he says, in *Burnt Norton*:

> Time present and time past
> Are both perhaps present in time future
> And time future contained in time past.

Tennessee Williams, in *The Glass Menagerie* (p. vii) relates to the Amerindian way of thinking in his comment that 'time is the longest distance between two places'.

Shakespeare revels in alternative conceptions of time. There are of course many instances of the standard collocations in his plays: people spend time, lose time, waste time, and so on, in the usual way. But in the nearly 1000 references that Shakespeare makes to 'time' we also find an extremely wide range of behavioural and mental metaphors, many of which in their personifications take us in the direction of the animating conception of time encountered in other cultures. 'A little time will melt her frozen thoughts', says the Duke in *Two Gentlemen of Verona* (III.ii.9). And in other plays we find time *untangling, reviving, sowing, blessing, conspiring, brawling, begetting, weeping, inviting, unfolding, ministering, expiring,* and doing much more. People also treat time in a much more innovative way: they *hoodwink* it, *redeem* it, *persecute* it, *confound* it, *greet* it, *name* it, *obey* it, *mock* it, *weigh* it, *jump over* it, and a great deal else. Indeed, in *As You Like It,* we find a dialogue between the lovers Rosalind and Orlando that turns our standard conception of time on its head. Rosalind is in disguise, and recognises Orlando, but he does not recognise her. She is feeling mischievous, so she tempts him into a word battle (*ibid.,* III.2.292ff).

> ROSALIND: I pray you, what is't o'clock?
>
> ORLANDO: You should ask me what time o'day; there's no clock in the forest.
>
> ROSALIND: Then there is no true lover in the forest; else sighing every minute and groaning every hour would detect the lazy foot of Time as well as a clock.
>
> ORLANDO: And why not the swift foot of Time? had not that been as proper?
>
> ROSALIND: By no means, sir. Time travels in diverse paces with divers persons. I'll tell you who Time ambles withal, who Time trots withal, who Time gallops withal, and who he stands still withal.
>
> ORLANDO: I prithee, who doth he trot withal?
>
> ROSALIND: Marry, he trots hard with a young maid between the contract of her marriage and the day it is solemnised; if the interim be but a se'nnight, Time's pace is so hard that it seems the length of seven year.
>
> ORLANDO: Who ambles Time withal?
>
> ROSALIND: With a priest that lacks Latin, and a rich man that hath not the gout: for the one sleeps easily because he cannot study, and the other lives merrily because he feels no pain; the one lacking the burden of lean and wasteful learning, the other knowing no burden of heavy tedious pedantry. These Time ambles withal.
>
> ORLANDO: Who doth he gallop withal?
>
> ROSALIND: With a thief to the gallows: for though he go as softly as foot can fall he thinks himself too soon there.

ORLANDO: Who stays it still withal?

ROSALIND: With lawyers in the vacation: for they sleep between term and term, and then they perceive not how Time moves.

Beaten, Orlando then changes the subject. This early instance of temporal relativity, anticipating Einsteinian insights by some 300 years, brings us closer to the way in which some cultures routinely think of time, as a relative, dynamic, influential, living force, and express it so in their verb forms, vocabulary, idiom, and figurative expression. Nor should we take the Einstein analogy too lightly, given that some theoretical physicists, notably F. David Peat in *Blackfoot Physics*, have actually attempted to work out the parallels between quantum physics and Amerindian practices and ways of thought.

I have only begun to give an account of the way English and other languages enable us to talk about time, and many more areas await investigation. For example, there is the question of how deaf sign language copes with the expression of time. As can be seen from Figure 5, with concept-based sign languages, such as British Sign Language, a line along the vertical plane, near the signer's ear and cheek, is regularly used to express time relationships. It seems a simple, unidimensional matter, with degrees of pastness

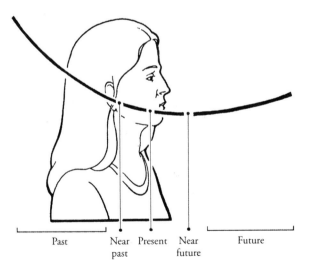

FIGURE 5. With concept-based sign languages, such as British Sign Language, a line along the vertical plane, near the signer's ear and cheek, is regularly used to express time relationships.

behind and degrees of futurity ahead. But do not be fooled by the unidimensional appearance of the diagram. For there are two sides of the head, and two hands to use, plus head movements and facial expressions, so that it is actually possible to do something in deaf sign language that is virtually impossible in speech and writing – express several points of time reference simultaneously. I once saw a signer talking about two people (let's call them A and B) who had read the same book, and she wanted to say that A had read it much more quickly than B. She expressed this by first identifying distinct areas of activity within the temporal space around the upper part of the body, assigning one area to A and the other to B. One hand then showed A beginning the reading in the near past and going on into the near future; the other showed B beginning in the more distant past and continuing into the more distant future. Both activities were signed at the same time. And the whole comparison took less than a second to convey. The treatment of time in sign language is plainly very different from what takes place in speech or writing.

I have not discussed the expression of time relationships in a discourse, either conversational or literary: how do authors vary the ways in which they talk about time? Nor has there been space to discuss the acquisition of time expressions by children: when do children begin to talk about time? Nor the issue of what happens when people lose control of their temporal expression, such as following a stroke, and find themselves unable to talk about time. But I hope that I have done enough to demonstrate the scope of this topic, and to illustrate its interest and challenge. Charles Lamb, in one of his letters (to T. Manning, 2 January 1810) remarked: 'Nothing puzzles me more than time and space; and yet nothing puzzles me less, as I never think about them.' We should think about these things, though, because apart from their intrinsic intellectual fascination there are all kinds of ready applications – to the teaching of language in school, where children are I fear still often taught misleading information about English tenses, or to work on intercultural understanding, where failure to realise that there are different conceptions of appropriate temporal behaviour can lead to communicative breakdown.

As Vladimir said, in Samuel Beckett's *Waiting for Godot* (Act I), 'That passed the time.' Estragon replies, 'It would have passed in any case.' 'Yes,' replies Vladimir, 'but not so rapidly.' I too hope that you have not been too conscious of time passing during this short excursus into (to adapt a term of Max Muller's) chrononomical linguistics.

FURTHER READING

Comrie, B., *Tense*, Cambridge: Cambridge University Press, 1985.

Crystal, D., *The Cambridge Encyclopedia of Language*, Cambridge: Cambridge University Press, 2nd edition, 1997.

Hall, E. T., *The Silent Language*, Greenwich, CT: Fawcett, 1959.

Palmer, F. R., *The English Verb*, London: Longman, 1974.

Peat, F. D., *Blackfoot Physics: A Journey into the Native American Universe*, London: Fourth Estate, 1995.

Quirk, R., Greenbaum, S., Leech, G. and Svartvik, J., *A Comprehensive Grammar of the English Language*, London: Longman, 1985.

Thieberger, N. and McGregor, W. (eds.), *Macquarie Aboriginal Words*, Sydney: Macquarie University, Macquarie Library, 1994.

7 Storytime and its futures

GILLIAN BEER

My title calls up childhood: 'Storytime' – a time set aside at home or in the school to listen to tales told, tales that are separated off from the busy routine of the day. That nursery setting is no accident. Listening to stories goes very deep down into personal history: it is there at the start of our learning where actuality and fantasy touch each other, and where they separate. It reinforces lines of descent. It tells us about the world before we were, and gives us a foothold in that world. Sylvia Townsend Warner, for example, dedicates her novel *The True Heart* (1929), with its submerged allusions to the tale of Cupid and Psyche:

> To my mother
> Who first told me a story

In stories and their telling, time is endlessly renewable. Those early storytimes when tales are told or read aloud give us warnings and promises about the future too, though the figures are at first sight improbable: grandmothers who turn into wolves, wolves who turn into mothers, glass slippers that do not cripple you.

But there is something concessive in the word storytime, something a little condescending, even drab. The time reserved for stories is both privileged and defensive: stories are (it is implied) different in nature from the ordinary and, if they seep into daily life, they do so with a warning. They are fugitives from the reservation: from that encircled *place* that is also a particular time. James Joyce evoked that in the opening of *A Portrait of the Artist as a Young Man*. The adult reader is dazzled anew, like a child, by the unannounced tale, its presentation implying quite a different relation between teller and listener from what we expect at the start of a novel, that adult form of desire. Yet the story is told by an adult in a special form of adult speech, baby-talk:

> Once upon a time and a very good time it was there was a moocow coming down along the road and this moocow that was coming down along the road met a nicens little boy named baby tuckoo... .
>
> His father told him that story: his father looked at him through a glass: he had a hairy face.

The child's recollection is plain, even peremptory; the father is the one infantilised.

I tell here two entwined stories about futures: about the reader's futures, bred, caught and shaped within the activity of fiction; about the future reader, unknown to the writer and only partly under the writer's control. Future perfect, the *Oxford English Dictionary* tells us, 'expresses an event or action viewed as past in relation to a given future time'. The reader's reading futures have more latitude than that suggests, more multiplicity, as we shall discover. Thinking about how we experience futures within fiction will allow me also to mark some distinctions between the novel and storytelling.

Quite recently there has been a striking return to the value of storytelling, both in its spoken and its written form. In pubs and arts centres formalised storytelling takes place, in which the oral is framed and timed. Among writers across Europe and beyond, storytelling is given renewed prominence. And whereas 20 years ago among academics the fashionable study was called narratology, now it is more likely to be story, as in John Niles's excellent study *Homo Narrans* (1999), with its emphasis on the remote pasts out of which stories have issued and on his provocative suggestion that stories never are quite oral, since we encounter them only through their having been written down. In narratology the study of time and tense *within* fictions, rather than their genealogy from the distant past, was emphasised. Indeed, the work of critics such as Gerard Genette offers one of the most subtle and lucid descriptions of how tenses control our reading destiny, how the time of *event* and the time of *recounting* interlock and strain to produce the tensile strength of fiction. But the aggrandising tendency in the term narratology (which after all simply means the study of narrative) has produced the counter-thrust that favours more concrete terms: storytelling or the tale. The debate is not new (Figure 1).

How to start the story? How to make time begin? How to breed futures? Those futures in story time are in the past, it seems, but we are to live them now – and to presage more of them than the story will confirm. Ciaran

LATE FROM THE SCHOOL-ROOM.

Minnie. "I AM READING SUCH A PRETTY TALE."
Governess. "YOU MUST SAY NARRATIVE, MINNIE—NOT TALE!"
Minnie. "YES, MA'AM; AND DO JUST LOOK AT MUFF, HOW HE'S WAGGING HIS NARRATIVE!"

FIGURE 1. Narrative or the tale? A mid 19th century cartoon from *Punch*.

Carson, Northern Irish writer and poet, opens his book *Fishing For Amber* (1999) like this:

> It was long ago, and long ago it was; and if I'd been there, I wouldn't be here now; if I were here, and then was now, I'd be an old storyteller, whose story might have been improved by time, could he remember it.

The reader/listener enters an incantation that giddies the sense of near and far, then and now. First the storyteller uncouples himself from the time of the story ('if I'd been there, I wouldn't be here now') – it was so long ago he'd be dead; or is there a hint too that if he'd been there he'd have been done away with by the events of the story? In the nature of many stories, as the saying recognises, you cannot act and '*live to tell the tale*'. So third person and past tense may be endemic to what is told. But, in any case, ('if I were here, and then was now, I'd be an old storyteller') even if *then* was *now* he'd be an old storyteller. In stories 'then' becomes 'now' for listener or reader. The story remains; the storyteller ages. The story is enacted in narrative time; the storyteller in natural time – the time of the body. The story ages only if the storyteller's memory goes: then it is lost. But even then it is lost only to *him*. He has told it many times. It has had many hearers. The story survives, changed by new telling, new circumstances, yet always received in the present, in the ear of the listener or the eye of the reader, into the interior. This particular storyteller claims he is not yet frayed by old age. (He is *not* the old storyteller.)

The passage continues:

> Three good points about stories: if told, they like to be heard; if heard, they like to be taken in; and if taken in, they like to be told. Three enemies of stories: endless talk, the clack of a mill, the ring of an anvil – I've often wished I could begin a story in the manner of my father, who liked to use a gambit like the one above to generate his stories told in Irish.

So (in a time recoil) this, it turns out, was his father's storytelling, in another tongue, not that of the writer now (though that of the writer now by annexation). Time and telling have slipped twice already: it is a typical storytelling paradox. Starting a story means disturbing here and now; it means invoking experience not yet known, which yet is experience that has already taken place within the story. The trick is how to signal that destabilisation – how to stabilise it. Often this is performed by incantation (Carson comments he cannot reproduce in English the balance of the Celtic sentences that invoke the new experience). Most often, the opening uses gambits and formulaic phrases to do with time: 'Es war einmal', 'Once upon a time'). An informant, Nadina Christopoulou, kindly tells me that in Romani folktales the formulation recognises also the nomadic nature of Romany social time: time and place change together. The habitual starting phrases are 'Yek drom' ('upon a road', or 'once upon a road'), or the oscillating presence/absence of 'Sine ta na sine' ('there was and there was not'), while for the ending the most common phrase is 'Ote sinom, akate avilom' ('there I was, here I came').

All these markers emphasise one essential thing about story. The story cannot take place right now, alongside. Actuality is fermenting; it includes too much. Actuality is too rich in sensory materials to be available instantaneously as narrative. As Thomas Hardy remarked in his *Journal* (27 January 1897):

> Today has length, breadth, thickness, colour, smell, voice. As soon as it becomes *yesterday* it is a thin layer among many layers, without substance, colour, or articulate sound.

The storyteller's or the writer's struggle is to evoke in narrative the richness of today through what is knowable only as yesterday. Of course, that richness of today, the everyday, is also humdrum. It becomes story when it is separated, articulated – when enough is *left out*, and time can move at different paces. Take this story, published in an anthology of stories by children, based on their own lives, *I'm Telling You!* (2000, p. 12): 'The Day I Walked in Wet Cement', by Bryan Clark (Figure 2).

FIGURE 2. 'The Day I Walked in Wet Cement'.

> When I was six, I was going to the shop, in Skegness. I stood in wet cement and I was there for a half an hour. By now the cement was drying and I got worried.
>
>   The shop assistant came out and tried to help me.
> It was no use. She could not get me out.
>
>   My dad was wondering where I was and he came over! My dad had an idea and he pulled my shoes off and I got out.
>
>   I thanked him all day.

'I was there for half an hour!' – utterly convincing and equally preposterous: the child's panicked experience and cool recollection combine. The spaces on the page nicely register time lapses and pure endless relief is expressed in that last sentence: 'I thanked him *all day*' (my italics).

  Any story has to start, certainly. There must be time for retrospection, too, or at the very least for the busy activity of writing things down. Indeed, a particular difficulty of first person is that the writer and the actor become entangled, as Henry Fielding brilliantly mimicked in *An Apology for the Life of Mrs Shamela Andrews* (1741), his take off of Samuel Richardson's *Pamela*, where Shamela is endlessly writing home in the most unlikely circumstances. In this passage, body and writing become hopelessly confused. Perhaps she is left-handed, as she scribbles away while warding off and deluding her master, Squire Booby:

> **Thursday Night, Twelve o'clock**
> Mrs Jervis and I are just in bed, and the door unlocked; if my master should come – Odsbobs! I hear him just coming in at the door. You see I write in the present tense, as Parson Williams says. Well, he is in bed between us, we both shamming a sleep; he steals his hand into my bosom, which I, as if in my sleep, press close to me with mine, and then pretend to awake. – I no sooner see him, but I scream out to Mrs Jervis, she feigns likewise but just to come to herself; we both begin, she to becall, and I to bescratch very liberally. After having made a pretty free use of my fingers, without any great regard to the parts I attacked, I counterfeit a swoon.

In his 1843 *Journals* Kierkegaard reflected:

> It is perfectly true, as philosophers say, that life must be understood backwards. But they forget the other proposition that it must be lived forwards. And if one thinks over that proposition it becomes more and more evident that life can never really be understood in time simply because at no particular moment can I find the particular resting place from which to understand it – backwards.

Or as the Queen tells Alice in Lewis Carroll's *Through the Looking Glass*, when describing the conditions she will offer Alice as a maid: 'Jam tomorrow, jam yesterday, but never jam today'. The joke would have sprung to life more rapidly for Carroll's first readers, who pronounced the joke-word as *jam*, where more recent speakers of Latin say *iam*. Iam or jam: the *now* of narrative, placeless and timeless: jam tomorrow and yesterday, but never jam today. To put it another way – fiction can freely traverse, re-assemble, and invoke past and future because it never is *taking place* (taking *space*, one might say) in the present. It is actual in the emotions, breeding potential outcomes in the reader's mind, but never located in any outside, right now.

So far I have taken my cue from story *telling*, as, by sleight of hand, the writer Ciaran Carson does (and many writers do). But listening and reading are very different experiences. Walter Benjamin, in his famous essay 'The Storyteller', with a certain nostalgia, summons up the figure of the storyteller in a room with his audience and sets it over against the novel (and perhaps particularly the modernist novel of the 1920s and 30s with its repudiation of narrative sequence). Benjamin emphasises the community of listening to a tale told now, warm from the speaker's mouth. The present time has special value here: at least two people, and very probably more, are in relationship at this moment, though what the tale tells may be drawn from the remotest past. The past tense of the teller is intercalated with the present tense of his or her being, being here and now, with the listener in the same space. The story told aloud is embedded in the real time of its telling, as well as the place of its events.

Benjamin suggests that the community on which such telling is founded can have no meaning in the written text, is even driven out by text. His essay has the quality of threnody, looking back upon a dying form of society. The intimacy of the oral draws its life from a particular kind of closeness, which in turn implies small spaces for living, meagre provisions, and body heat. Affirming the presence of both teller and listener has to do with survival – with being alive together against the odds. This accounts for the heroic tone of Benjamin's essay, which is never quite directly addressed by its argument. As a result, written fiction for him is viewed with suspicion, bourgeois in its freedom from the heat and cold of a shared and intimate space. The reader reads silently, alone.

It is indeed true that once we move from storytelling to novel or romance the reader's time and space is changed utterly, as is the temporal relationship between writer and reader. Now the reader may be anyone who knows the language, at any point in the future. No speaking body is needed, no shared social circumstances. From the point of view of the writer the anonymity of the possible (and multiple) reader is indeed a daunting freedom. But for the reader it has many advantages, not least that the magical duress of the storyteller recedes.

Now the reader inhabits multiple time at will. No longer can the taleteller control and conjure response by raising or lowering the voice, by referring to things in the room, or to current happenings. The close fug of community is dissipated; the cool air of the written allows far greater control to the reader, first, in the timing of reading. The reader can pick up the book and put it down, can expand the time of reception, can meld thinking about the fiction with everyday occupations in the intervals between reading, can flick through to glean a sense of what is coming, or even enjoy the illicit delights of reading the last page beforehand (a delight that carries disappointment at its core). The reader can read backwards, even sideways, as well as forwards, as Virginia Woolf describes Mrs Ramsay doing in *To the Lighthouse* (1928): 'she read and turned the page, swinging herself, zigzagging this way and that, from one line to another as from one branch to another'.

The storyteller is all powerful in the group; the writer, on the other hand, must collaborate with unknown others, each reader living in future time. The storyteller's power lies in the occasion, in *his or her now*; the writer's must lie in the future, in *our* now. The writer therefore must transfer onto the page many of the cues and hints held in voice and gesture and re-awaken intimacy as a form of thought – the *reader's* thought. The writer presages event, precipitates plot, allures us forward with suggested outcomes, allows us to ruminate perhaps on the potentialities of different characters. But the reader draws into the story circumstances that the writer, locked in the past of writing, cannot control. The reader also cranes forward through the fiction, imagining alternative futures at every point.

The subject of fiction is most often passion, in its many manifestations, as sexual desire, desire for knowledge, for gold, for dominance or survival, and for renewal, for what has been lost, for return to origins; its medium is anxiety. Both passion and anxiety are forms of the future, though they may take their images from the past. In a long novel the reader must need to read on, to turn the page, to scan and trawl for clues as the lover does. The tug of the story is towards ending and the novelist and fiction writer conjure an extraordinary array of devices to thwart and delay conclusion (and at the same time satisfy the reader). Dalliance and resistance build up plot. The reader must want to know the future, but one of the principal pleasures of reading is the reader's power of imagining multiple alternative outcomes at every moment of the text. The writer spreads out a fan of possibilities. The reader hypothesises the turns of event and feeling. We are pleased to be disappointed as well as pleased to have our hypothesis confirmed. That pleasure depends on fiction's constant tribute to the alternative futures loaded into the reader's imagination. Curiously, these alternative futures can survive even into a second reading, seamed into syntax, making us suffer anew Emma's errors in Jane Austen's novel of that name (though now we see them coming) – or enduring afresh Emma Bovary's misunderstandings as Flaubert describes her attempts to grid romantic fiction onto provincial life. We *still* foresee and hope for alternative outcomes at every point in our reading.

The extraordinary power of Samuel Richardson's *Clarissa* (1747), perhaps the founding text of the European novel after Cervantes's *Don Quixote* (1605 and 1615), derives in large measure from the way the reader is provoked to mimic in the process of reading the novel's wayward yet inexorable movement towards the rape of Clarissa. The length of delay, the intricacy of various lies and fantasies generated by Lovelace in his pursuit of her, produce titillation and exhaustion in the reader so that, against all judgement, we begin – *as readers* – to want the climax to take place. The book puts the future in jeopardy and the reader joins in the obsessional pursuit of that one event to come, joins equally in the twists and turns by which, together with Clarissa, we seek mentally to avoid it. Despite its thronging multiplicity, its 'writing to the moment' through letters from many different people, what is displayed in this immense work is the speed with which the event is re-written and amended, in immediate memory and in record. It displays also the voraciousness of readers. The work bulges with manifold futures conjured into life by its readers (encouraged by its author). *Now* we know its outcome in advance. Its first readers, as it appeared in instalments, were begging Richardson to deviate from the narrative's dire impulsion and to save Clarissa before it was too late; that is, before her fate was written. (For example, Lady

Bradshaigh, writing as Belfour, wrote to Richardson: 'Dear Sir, if it be possible – yet, recall the dreadful sentence; bring it as near as you please, but prevent it. Do, dear Sir, it is too shocking and barbarous a story for publication. I wish I could not think of it. Blot out but one night, and the villainous laudanum, and all may be well again.') Many bolt-holes were offered by the narrative; all were blocked. That closing down of lives to come is a powerful and painful effect in fiction. It relies upon fiction's power of generating plural futures in the reader's mind, against the grain of knowing that the text is already written. That temporal paradox is central to the joy and grief of the reading experience.

Reading fiction allows us to survive it. The book closes. We are still there, without having to live out all the consequences experienced within the measure of the work. In that sense, like drama, the time of the story allows affirmation. We continue, into other daily futures. But, as I have just suggested in speaking of the immensely long novel *Clarissa*, fiction also teaches us determination (only certain events out of a number of future possibilities generated actually happen in the story) and termination (ending). Some releases do not occur. Lives end. Books end. Overdetermination haunts the text. Multiple times combine inexorably.

If *Clarissa* demonstrates that, so does Kafka. His 'Little Fable' reads like a nursery tale. Its outcome is even more dire than 'Three Blind Mice', partly because it includes more time zones and tenses.

> 'Alas,' said the mouse, 'the world is growing smaller every day. At the beginning it was so big that I was afraid, I kept running and running, and I was glad when at last I saw walls far away to the right and left, but these long walls have narrowed so quickly that I am in the last chamber already, and there in the corner stands the trap that I must run into.' 'You only need to change your direction,' said the cat, and ate it up.

The five lines encompass the sweep of a life's experience. The baffling largeness of the child's world narrows into adult containment and then constriction, enclosure, premature ending. Two voices speak, in the present, though not quite in dialogue. The laconic riddle of the cat's advice allows no reply from the mouse. The mouse has a choice; a different future is possible, despite the desperate architecture of the corner and the trap. The swerve (does he swerve? – that is moot) brings immediate death. The trap is sprung on the reader. The whole is held in the narrative past tense: said the mouse, said the cat, the one action: 'ate it up'. Intriguingly the English here has a nursery turn: 'ate it up' suggests childhood injunctions. The original German uses the term for animals eating: *fressen*, not humans, *essen*. This is a fable. It repeats an action endlessly. It hints at inexorable meaning. It does not encourage free interpretation. We move from infancy recollected as panic through to a time that allows no recollection at all. The mouse is dead. Here is no repartee. Punctuation in the English version gives a moment's interruption and emphasis to the ending (comma, 'and ate it up'). In the German, speed and seamless consequence are emphasised: the cat's motion is mimicked in the continuity of speech and action, its laconic closing down. Time-telling even within these two extremely brief texts is subtly varied by language and also by punctuation systems, those often unobserved but crucial elements in meaning.

> 'Ach' sagte die Maus, 'die Welt wird enger mit jedem Tag. Zuerst war sie so breit, dass ich Angst hatte, ich lief weiter und war glücklich, dass ich endlich rechts und links in der Ferne Mauern sah, aber diese langen Mauern eilen so schnell aufeinander zu, dass ich schon im letzten Zimmer bin, und dort im Winkel steht die Falle, in die ich laufe.' 'Du musst nur die Laufrichtung ändern', sagte die Katze und frass sie.

The novel proffers novelty, the new. The reader enters. Will we experience the mouse's terror, trapped in an inevitable corner, overdetermined? Or is this the largesse of multiple futures, the free resurrections of the lost, as in romance or in Jane Austen's *Persuasion*, that most moving of ghost stories, in which past love, past beauty, past youth almost impossibly return and Captain Wentworth, revenant, opens out a wholesome future that we do not need to have described, so inward are we with Anne Elliot by the last pages of the novel.

In *Beginnings* (1977), Edward Said remarked that beginning includes within it the intention to continue. A common tease-structure for nursery story plays frustratingly on that promise of time to come.

> I'll tell you a story about Jackanory.
> Now my story's begun.
> I'll tell you another about his brother.
> Now my story's done.

The rhyme summarily snaps shut the empty box. Begun is done within a single breath. The story is not so much truncated as extruded. (If it's happening at all it's elsewhere, nowhere we'll ever be.) The more threatening Gothic version of this technique (and there are many such openings) begins:

> It was a dark dark stormy night and the Captain said 'Tell us a story'.
> 'It was a dark dark stormy night... .'

Hysteria lies in wait for the listeners to this unstoppable round. Both these infantile forms get their zest from the drive to continue, onward. The listeners, or readers, are thwarted in their appetite for the future. The narrative has gobbled its own tail, to its own satisfaction, but not to ours. This continual beginning has been taken to a florid extreme in Italo Calvino's novel *If on a Winter's Night a Traveller* (1980), which plays across the nature of storytelling and novel writing. For once, the claim is that the author is speaking (writing) now, alongside the reader. Breaking into the free anonymity of the novel-reader, secluded in the future of the writer and the written, the narrator instead uses second person 'you', and tells us authoritatively what we are about to do:

> You are about to begin reading Italo Calvino's new novel *If on a winter's night a traveller*. Relax. Concentrate. Dispel every other thought. Let the world around you fade. Best to close the door; the TV is always on in the next room.

Maddeningly he busies himself about our comfort, for his own convenience. Once in, we are to have no out he warns us. Total control:

> Adjust the light so you won't strain your eyes. Do it now, because once you are absorbed in reading there will be no budging you. Make sure the page isn't in shadow, a clotting of black letters on a gray background, uniform as a pack of mice; but be careful that the light cast on it isn't too strong, doesn't glare on the cruel white of the paper, gnawing at the shadows of the letters as in a southern noonday. Try to foresee everything that might make you interrupt your reading. Cigarettes within reach, if you smoke, and the ashtray. Anything else? Do you have to pee? All right, you know best.

All this preparation promises an engrossing read without interruption. First preening himself and the reader, then soberly, the author seems to promise a full satisfaction for our curiosity, a satisfaction that will depend, not on him, but on the coming tale:

> it's the book in itself that arouses your curiosity; in fact, on sober reflection, you prefer it this way, confronting something and not quite knowing yet what it is.

A modest heroism is flatteringly assigned to the reader: 'I prefer it this way, confronting something and not quite knowing yet what it is.' Our stamina and experience is affirmed. A knowable future is about to be explored. But the joke will be on us: our confidence in our own readerly powers to guess at futures is about to be tested, and flouted. In the event, the tale is one of frustration, a succession of first chapters in quite different styles, from apparently different books, which yet together make our experience of reading in this novel. Calvino hedges his narrative bets by providing a hopeless love pursuit of the *other* reader (here female) whose experience no current reader ever can quite know. That story gives a thread of open (perhaps sentimental) future that makes tolerable the recursiveness of the book's strategy. The narrative trap of constant return to a new beginning is both exhilarating and exasperating. It saps passion and exaggerates anxiety. The work remains a *jeu d'esprit* that teaches us in a somewhat pedagogic mood all about folktale and written text, as well as engaging us in an evening's game.

But the embedding of foreknown story in wayward consciousness can become part of a profound exploration of how we survive and how we die, as readers living the future perfect, practising life, in both senses. Two of the richest 20th-century novels turn to the same Brothers Grimm tale to feed their narratives. Folktale in each of these novels provides a ballast of the past for the reader; we are assumed to know the tale already. Events recounted as new in the sequence of the story are already embedded in our past. But the novelist then breaks the boundaries of the foreknown and implicates the story in fresh and unforeseeable experiences. Günter Grass's big novel *The Flounder* (1977) centres on the edible talking fish from what the narrator calls the "misogynistic fairy tale 'The Fisherman and his Wife' with its parallels as far afield as Africa and India". Grass's book is full of disastrous history, poetry, fierce class and gender debate, and above all food. The recipes, for 'white beans cooked down to a purée served with roast pork and pepper sauce' (p. 205), 'sheep's-milk cheese with smoked cod liver' (p. 83), 'manna grits with sorrel' (p. 168), have the fierce flavours of those who have known hunger, what Esau in one of the book's poems calls 'lentil law' (p. 190). The reader joins in conjuring appetite: the food lies always just beyond the mind's reach, about to be tasted, warm and heavy in the head, evoking repletion rather than flavour and inclining always onwards towards detritus and death: 'So much love, ready for the dustbin' (*ibid.*, p. 168).

Storytime and its Futures

> When Agnes the cook
> kissed Opitz the dying poet,
> he took a little asparagus tip with him on his last journey
>
> (ibid., p. 234).

Food and its consumption is, in the time of this fiction, a constant succession of little lives, little deaths, evoked but never consummated. Eating is survival. In appetite the future is always renewed and squandered. The reader strains towards bodies and stories we never will quite share. Yet Grass's novel is organised through the nine months of a pregnancy, a biological time that we can neither abbreviate nor prolong. And at the end the husband-narrator presages a new world for women in which the Flounder, that magic talking fish (Figure 3),

> the flat, age-old, dark, wrinkled, pebbly-skinned Flounder, no, my Flounder no longer, leaped as though brand-new out of the sea and into her arms.
>
> I sat beside the empty dinner pail. Fallen out of history. With an aftertaste of pork and cabbage.

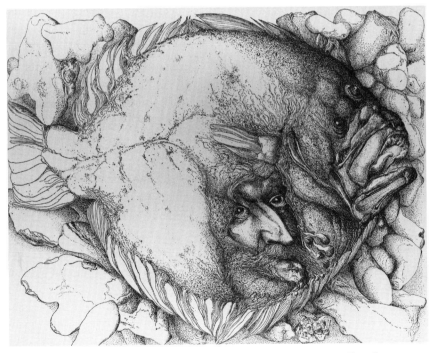

FIGURE 3. Günter Grass's own visualisation of the eponymous Flounder.

This time the fisherman's wife has got her way, it seems.

Virginia Woolf's *To the Lighthouse* threads the Grimms' tale of 'The Fisherman and his Wife' through the first half of her novel. The tale is, as in Grass's novel, about the importunate wife who drives her husband again down to the sea to beg a wish from the magic fish, the flounder. Her desires and her rancours expand each time the wish is granted. The Grimms' tale is also a tale about seeking security: bringing things to perfection, getting control. Woolf is exploring that desire as well.

> She read on: 'Ah, wife,' said the man, 'why should we be king? I do not want to be King.' 'Well,' said the wife, 'if you won't be King, I will; go to the Flounder, for I will be King.'

The mother herself reads aloud the tale to her little boy James, only half-hearing it as she does so and half exploring the furthest reaches of her consciousness, making up alternative and parallel stories of her young friends' futures. She is not thinking about the story of the fisherman's wife. She is waiting for them to come back from the sea-shore; it is growing dark and late. Has Paul asked Minta to marry him or not? She broods, and we with her, on these immediate alternating futures of yes and no, and beyond them on her own being and behaviour, her hospitality, her allure.

> However, Minta came.... Yes, she came, Mrs Ramsay thought, suspecting some thorn in the tangle of this thought; and disengaging it found it to be this: a woman had once accused her of 'robbing her of her daughter's affections'; something Mrs Doyle had said made her remember that charge again. Wishing to dominate, wishing to interfere, making people do what she wished – that was the charge against her, and she thought it most unjust.

Is she domineering? Does she seek to own others? We enter a saturated scene of brooding where the Grimms' tale is lost yet present. To read those pages aloud would change radically the enactment offered to the reader. The silent words on the page move, wrapt, through that familiar level of frank self-acknowledgement that lies in our own minds just beneath utterance and must remain unuttered. The reader, with Mrs Ramsay, is trying out imaginations of what might be, has been, inexorably is and is to be, almost as if she and we were one, alive in a common present. Save that the story of the fisherman's wife lies lodged in us, too: cranky, observant, and fugitive.

The silent space of the reader's mind performs the motions of a consciousness, even as it plays doubtingly across them. The identification and the slight thrill of dissent demands extended silence. The experience here of fused present, past, and future relies on intertwining spoken story and silent consciousness. Its effects are achievable only in the novel, not in story as Benjamin imagined it, nor indeed as it is performed by tellers of tales. It brings out the complex dependencies between spoken and written words in fiction, and between repeated pasts and futures that we never can enough foresee, though so much of fiction's pleasure is in imagining their variety. At last Mrs Ramsay's outer task is nearing its end. The story is almost over and its formulaic ending enacts the paradox of fictional time: gone and enduring, speaking still but shutting the book. Mrs Ramsay is anxious for her young: 'one of them might slip. He would roll and then crash. It was growing quite dark.'

> But she did not let her voice change in the least as she finished the story, and added, shutting her book, and speaking the last words as if she had made them up herself, looking into James's eyes: 'And there they are living still at this very time.'
> 'And that's the end,' she said [...].

Ending. Survival. We read in the unknown future, a future in part generated by past writing, making fresh meaning out of stories composed before our time, lived through in our present, and surviving past that present. When we recollect novels or stories, they do not return to us played out in the formal time sequence of plot. The whole has resolved to images, or to a special temper of feeling, tesselated not only from what *happened* in the tale but from its possibilities, impressions, its futures seized as experience.

FURTHER READING

Barchas, J., *The Annotations in Lady Bradshaigh's Copy of Clarissa*, Victoria, British Columbia: English Literary Studies, no. 76, 1998.

David, C., Lenoir F. and de Tonnac, J.-P. (eds.), *Conversations About the End of Time: Umberto Eco, Stephen Jay Gould, Jean-Claude Carrière, Jean Delumeau*, trans. Ian Maclean and Roger Pearson, London: Allen Lane, 1999.

Eco, U. and Sebeok, T. (eds.), *The Sign of Three: Dupin, Holmes, Peirce*, Bloomington, IN: Indiana University Press, 1983.

Genette, G., *Narrative Discourse*, Oxford: Basil Blackwell, 1986.

I'm Telling You! Compilation from the 1999 Cambridge Young Writers Award, with a foreword by David Blunkett MP, Cambridge: Cambridge University Press, 2000.

Niles, J. D., *Homo Narrans: The Poetics and Anthropology of Oral Literature*, Philadelphia: University of Pennsylvania Press, 1999.

Rubin, D. C., *Memory in Oral Traditions: The Cognitive Psychology of Epic, Ballads, and Counting-Out Rhymes*, Oxford: Oxford Uuniversity Press, 1995.

Said, E., *Beginnings: Intention and Methods*, New York: Basic Books, 1975.

8 Time and Religion

J. R. LUCAS

Introduction

Religion does not take kindly to time. For many people the impulse to religion is generated by a desire to escape from the tyranny of time: we seek a refuge from the changes and chances of this fleeting world, and hope to evade the inevitable mortality of our temporal existence, by moving to a higher plane of reality, where the limitations of temporality are transcended by the eternal verities of absolute existence. And, independently of our motives for seeking God, it would seem to derogate from His perfection to subject the Almighty to the corrosion of time. So the Greeks, once they began to emancipate themselves from the all-too-human gods of Olympus, started to posit a timeless Absolute, an impassible, unmoved mover, the ground of our being, and perhaps the worthy recipient of our worship, but not an active intervener in our affairs or a person with whom we could communicate

The God of the philosophers was clearly very different not only from the Olympian deities but equally from the Yaweh of the Jews. The God of Abraham, Isaac and Jacob was thought to have intervened on many occasions in the course of the history of Israel, giving them a helping hand in their escape from Egypt, and a chastening one when they went after false gods. The word of the Lord came to the prophets, often unwelcomely as it did also to David and Ahab when they strayed from the straight and narrow path of righteousness. Jesus constantly addressed God as *Abba*, Father, and the Christian understanding was based on God's having decisively intervened in sending His Son into the world and in having raised Him from the dead, something that a Parmenidean God would never have stooped to do.

There was an obvious tension. Tertullian recognised it: 'What has Jerusalem to do with Athens?', he rhetorically asked (Figure 1). But most people

Yaweh	**The God of the Philosophers**
Jesus	**Parmenides**
Anthropomorphic	**Absolutely Absolute**
	Source of Existence
Revelation	**Theological Superlative**
Personal	**Impersonal**
Temporal	**Timeless**
Does things	**Ground of our Being**
led His people across Red Sea	
raised Jesus from the dead	
Has feelings	**Impassible**
'40 years long was I grieved with this generation'	
'unto whom I sware in my wrath'	
Can change mind	**Changeless**
'Father, if thou be willing, remove this cup from me'	
Limited foreknowledge	**Omniscient**
Vulnerable	**Perfect**

FIGURE 1. 'What has Jerusalem to do with Athens?' – Tertullian.

preferred not to be very much aware of the tension. Philo sought to accommodate Judaism to Greek philosophy by understanding most of the Old Testament mythologically, much as modern believers reconcile Genesis with geology or cosmology by insisting that the account of creation is not to be taken literally but expresses a deeper, metaphorical truth. Justin Martyr sought to make Christianity philosophically respectable, and his efforts were followed by a concerted endeavour on the part of the Fathers in late antiquity to achieve an intellectually acceptable understanding of Christianity. God became more and more respectable, and less and less human. In the contest between Athens and Jerusalem, Athens won.

I want to challenge that result. The God of the philosophers cannot do the work the Christian God is required to do. I shall argue that, in spite of great ingenuity on the part of Augustine and Aquinas, we do not have an intelligible account of how a timeless Being can act in history; in particular, how a timeless Being can, like a father, pity His children, hear their prayers, and on occasion respond to their petitions. In this chapter, therefore, I shall first unravel the different considerations that have led thinkers to hold that God must be timeless; I shall then argue that that is to depersonalise Him; and finally I shall try to sketch a more positive – and temporal – view of eternity.

Why suppose God to be timeless?

We suppose God to be timeless for many reasons. Religious experiences seem out of this world, and so, we think, outside time too. Thinking about the God of the philosophers is an exercise in the theological superlative, which all too easily oversteps the bounds of common sense and even intelligibility. And if God is timeless, we can fend off awkward questions – about foreknowledge and free will, and about the beginning and end of the universe – which otherwise would be difficult to answer. Underlying these considerations are certain mistakes – about the nature of change, about time's similarity with space, and a confusion between instants and intervals. Once these are recognised, we no longer feel impelled to think that God must be timeless.

Time implies the possibility of change, and Plato was against change. We can sympathise. It is natural, when change rears its ugly head, to want to stop it, not only for the immediate now, but for ever in principle. If reality *is* rather than *becomes*, then we can discount unwelcome changes as mere transitory appearances, and not permanently real. Many truths, for instance those of mathematics and the natural sciences, are either timeless or omnitemporal, holding at all times, as they do also in all places and for all persons. Just as the laws of nature are invariant over time, so the Ultimate Reality must be changeless and free from any temporal variation.

Plato had a further, more explicit, argument for the changelessness of God. If God changed from one state to another, then either the change was for the worse, in which case God ended up less good than He would have been if He had not changed, or the change was for the better, in which case God's previous state was less good than it might have been. In either case God would be less than perfect. The perfection of God required that He be changelessly at the acme of perfection. But this argument assumes that there is a strict linear ordering of states with respect to moral merit, and this is not obviously true. Many changes – my breathing in and my breathing out, for example – are matters of indifference so far as moral merit is concerned, and even where moral virtues are concerned, they are not always either compatible or comparable: I cannot be tactful and understanding and at the same time courageously standing up for the right and the good. It is the same as with aesthetic merit: we hesitate to say that either the Parthenon is more beautiful than Santa Sophia, or Santa Sophia is more beautiful than the

Parthenon, and do not feel impelled to hold either that Bach is better than Beethoven or vice versa. There does not have to be a single uniform order of over-all merit and, once we recognise this, Plato's argument from perfection falls to the ground.

More fundamentally, Plato has misconstrued the logic of change. When we talk of change, we need to be able to answer the question 'Change with respect to what?'. At a reunion I come across old Bloggs, and exclaim 'Why! You have not changed one bit'. I mean that he has still the same characteristics, is still telling the same funny stories, making the same, rather feeble puns. But, of course, he has changed in other respects: he is fatter, balder, richer than he was when we had adjacent rooms in college. Indeed, he would not be able to tell those funny stories if his lips did not move and his lungs breathe in and out. It is quite reasonable for Plato and the psalmist to want God to be unchangingly reliable and faithful, but it does not follow that He must therefore be absolutely changeless (Figure 2).

Theists are easily confused when they come to consider the relationship between God and time. If God exists, He is the ultimate explanation, the ultimate be*cause*, He is the first cause, the creator, the maker of all things. He

---

What is Change?

Something changes if it is **Different** at a different time.

What is the logic of the word 'different'?
Different is a **3-term** relation:

$$x \text{ is different from } y \text{ in respect of } \textbf{A-ness}$$

(I am different from **you** in respect of **height**, but not in respect of **being human**, or **being interested in time**.)

It is natural to want God to be changeless in the sense of being always reliable, not capricious, trustworthy, not fickle, but quite unreasonable to require him to be unmoved by our plight, unresponsive to our prayers.

The Absolutely Changeless is unlikely to do its metaphysical work: in classical materialism the atoms, although always the same in themselves, have different positions and velocities at different times.

FIGURE 2. Change.

therefore made time, and so cannot be in time, since He existed before time was. But is time the sort of thing that can be made? If God made the universe, we can ask 'When did He make it?', and answer 'About 1.5×10^{10} BC'. But once we see that time is different from change and things that can change, we can no longer ask 'When did God *make* time?'. Many philosophers have been reluctant to allow this, because they think it downgrades God to being a mere Demiurge, a glorified Lord Nuffield, who manufactures things within a pre-existing time and space that constrain His activities. But what is being denied is not the ultimacy of God but that the relation of God and time is one of *making*. God is not the mechanical *cause* of time, but He is the explanatory be*cause*. Time was *not made by* God but *stems from* God as being a personal Being.

Can God be timeless?

Theists are similarly confused when they ask themselves 'Is God in time?', and feel obliged to answer 'No', for fear of making God subject to time. We do not locate God in space, for that would be to circumscribe and limit Him. God is outside space, we are inclined to say; we can well imagine ourselves having a non-spatial experience – perhaps listening to the music of the spheres, perhaps in deep conversations with the saints who have gone before. Boethius puts it well in his *De Trinitate*: God is not present in any place, but every place is present to God. Hence, it would seem by parity of reasoning, God is outside time too, and when we are with Him, our experience will be timeless too (Figure 3). But time is not like space. When I first tried to think about the two, I concluded that, though space was like time, time was not like space; I was careful to call my book *A Treatise on Time and Space*, and not *on Space and Time*. The spatial analogy is misleading, and does nothing to show that God is timeless. Although God is not located within time, He is not outside it either, since every time is a time when He exists.

God's time has come under attack in the 20th century from a quite different quarter. Einstein and Minkowski seem to have justified Boethius: time is on a par with space. Instead of the absolute distinction, presupposed by common sense and articulated by Newton, we should regard spacetime as the underlying reality, with time and space being merely perspectival effects depending on our point of view. But time is not just a fourth dimension of space.

> **Is God in time?**
> Awkward to say he is: would he not then be enclosed in time, subject to time?
>
> Compare question: **Is God in space**?
> No; God is not in space: all space is present to God, space is the *sensorium* of God, all space is in God, God is outside space.
>
> *De Deo vero non ita, nam quod ubique est ita dici videtur non quod in omni sit loco (omnino enim in loco esse non potest) sed quod omnis locus adsit ad eum capiendum, cum ipse non suscipiatur in loco.*
>
> Boethius, De Trinitate, 4
>
> **Eodem modo** Boethius, God is outside time.
>
> But: **time is not like space**.
>
> **Although God is not confined within time, He is not outside it either, since every time is a time when He exists.**

FIGURE 3. Is God in time?

In the Special Theory, which is essentially a theory of electromagnetism, the fundamental manifold is spacetime, not space, and it is not just a four-dimensional analogue of our familiar three-dimensional space, but a (3 + 1)-dimensional manifold with a Lorentz signature. The difference between time-like and space-like dimensions and between time-like and space-like separations is profound, and the light-cone topology of spacetime is quite different from that of a normal Euclidean space, however many dimensions the space may have.

The Special Theory of Relativity seems to make the distinction between past, present and future relative to frames of reference. In one frame of reference I shall ascribe a future date to some distant event, while in another I shall ascribe a past date. This is a much more serious objection. If God exists, and can know of events happening, and whether one event happened before some other event, then there is some divine criterion of simultaneity, defining some absolute frame of reference, so that not all inertial frames of reference are equally good. That contravenes the Principle of Relativity as commonly expounded, but is not the scientific blasphemy it is often taken to be. We can best see this if we meditate on Newtonian mechanics.

Newtonian mechanics was also relativistic. All inertial frames of reference were equally good. No rest frame in space could be identified by mechanical means alone. But this did not mean that none could exist, or that absolute space was a meaningless concept. Newton argued that it was meaningful, and opined that the centre of mass of the solar system might be at rest in it. The fact that it could not be identified by mechanical means did not mean that it could not be identified at all. Consider the possibility that Newton's other researches might have been successful, and that the book of Ezekiel, properly understood, gave guidance on the matter. Had Newton been able to plumb the depths of Old Testament hermeneutics to identify those frames of reference that were truly at rest, he would not have hesitated to accept its findings. Nor were physicists obliged to resort to theology. Physics itself gave grounds for identifying an absolute frame of reference. That was the original reason for the Michelson–Morley experiment to measure the earth's velocity through the ether. Although an absolute frame of reference could not be identified by purely mechanical means, once physics extended beyond mechanics, it became possible that some further considerations might enable us to pick out an absolute frame of reference. The conjunction of Newtonian mechanics with electromagnetism gave rise to the ether, which could plausibly be regarded as being at absolute rest. In the event the ether was not discovered, and Newtonian mechanics was modified to bring it into line with electromagnetic theory. But if the ether had been discovered, it would not have shown Newtonian mechanics to be wrong, but merely vindicated Newton in showing that an absolute frame of reference absolutely at rest, though not required by Newtonian mechanics, was none the less consistent with it. So, too, now, if divine omniscience gives rise to hyperplanes of absolute simultaneity, there will be no inconsistency with the Special Theory. The Principle of Relativity will still hold: it will hold within the Special Theory, which is a theory of electromagnetism about electromagnetic phenomena. If we want to deal with the emission of photons in distant places, or the reception of wireless messages, or the distance between atoms in a molecule bound together by electric forces, then the best way to harmonise all our data into a coherent whole is to ascribe to distant events the dates given by the Lorentz transformations. All the equations expressing electromagnetic laws are covariant under those transformations, just as all those expressing the laws of Newtonian mechanics are covariant under the Galilean transformations.

Many scientists will be unpersuaded. For them it would be a blasphemy to allow science to be contaminated by theology, and if Newton was a theologian, it is something to keep quiet about, the later infirmity of what had been a noble scientific mind. But it is not just the theologians who want to restrict the Principle of Relativity. Many workers in General Relativity posit a cosmic time and a preferred frame of reference, and on any realist construal of quantum mechanics, there is a matter of fact about the time when Schrödinger's cat actually dies, and whether it is simultaneous with, before, or after, some distant event, say the absorption of a photon by a sodium atom in Alpha Centauri. If absolute time be anathema, let General Relativity and quantum mechanics be anathematised too.

Minds and time

Assimilating time to space was a mistake, but it was not just a mistake. It was supported by further features of our thinking about time. When we think not about time as it appears in physics and natural processes but about our way of thinking of it, we notice that we can choose our temporal standpoint. I can choose to think of Cambridge, not as it is now but as it was in my youth, or as it was when Newton or Darwin were in residence, in something like the way that I can see things not only from my own point of view, but from yours, or from that of some historical figure. I can conjugate over tenses, as I do over persons, and can, to a limited extent, free myself from the here and now, as I can, to a very limited extent, from the egocentricity of me and mine (Figure 4). In our philosophical moments, we try to extend our thoughts to think about the whole of time; then often lose our grasp on its essential temporality through a confusion between instants and intervals and a misunderstanding of what it is to be present – itself a word that can be used in either a spatial or a temporal sense. We think it essential to time that there should be a present – if we de-locate events from our temporal sequence, we deny that they ever really happened: 'once upon a time' is our way of indicating a fiction. And together with a present there must also be a past and a future with which to contrast the present. But a present what? A present instant? Or a present interval? Augustine puts forward an argument – I call it the argument of the ever-shrinking present – to throw doubt on the objective existence of time. Although we talk of the present, we find when

> Unlike material objects, minds can be in two places at once:
> I can be in Cambridge, but also imagine myself in Timbuktoo.
> So too, I can project myself into your shoes, and see things from your point of view.
> I can project myself into Wellington's mind, and think how to counter Napoleon's manoeuvres in June 1815.
>
> So too again, I can project myself into other times, and think of things from a different temporal standpoint from the one I now occupy.
> I can project myself back to June 1815, and think, without benefit of hindsight, of the situation in Brussels then.
> Or I think myself forward to 2035, and think of the recriminations that will be made then on account of our failure now to prevent global warming.
>
> Unlike material objects, minds can choose to adopt a different temporal standpoint from the one that they currently occupy.
>
> Philosophers are natural 'egotheists'. Each thinks of himself as God [Very dangerous in political philosophy] and readily takes it upon himself to be a spectator of all time.

FIGURE 4. Minds and Times.

we think about it that whatever interval we had called the present is not really all present but is partly future and partly past. At the time of writing, the year is 2000 AD, but January and February are already past and April is yet to come. And similarly with months, weeks, days, hours, minutes and seconds, until we seem forced to the conclusion that there is no time that is truly present, and so no real time at all. But what we have actually shown is that what counts as the present interval depends on context, and contexts vary (Figure 5). There is no absolute present interval, but rather a sequence of nested intervals that converge on an instant, the present instant (Figure 6), much as Cantor constructed a Cauchy sequence of nested intervals, converging to a punctiform limit, in his definition of a real number.

Plato relies on the converse notion – that of the ever-expanding present – in his characterisation of the philosopher as the spectator of all time. He is supported by our ability to adopt different temporal standpoints, and can think of things as we would have seen them had we been seeing them then, in much the same way as we can project ourselves into others' minds and

> The present is an **instant** of no duration dividing the future interval from the past interval.
>
> Past   Present   Future
> -------------------- | +
>
> The present is an **interval** between the future interval and the past interval.
>
> Past   Present   Future
> --------------- * * * * * * * * * * * * * * + + + + + + + + + + + +
>
> Question: **How long is the present interval?**
>
> Answer: **It depends:**
> present hour, present week, present term, present year, present century.

FIGURE 5. Present instant and present interval.

see things from their point of view. Here, in this series, it is an exercise of piety to think about Charles Darwin, and use appropriate tenses to refer to things from some time in his life, and say that when Darwin *was* on the Beagle, he *had already won* a good reputation in Cambridge, or that when Darwin *was* on the Beagle, he was thinking more about birds than girls. So, too, a philosopher can think about the Big Bang and what it would have been like to have been around then, and we can wonder whether or not there will be a Big Crunch, and envisage the universe either being cremated in a final implosion or subsiding without even a whimper into an uneventful heat death. When we survey the whole of time (and the whole of existence), our own life seems no longer to be of great consequence, and we can easily suppose that, since all time is present to us, none of it is past or future, and therefore it is not really time, since time is essentially a passage from future potentiality through present actuality to past immutability (Figure 7).

Both the argument of the ever-shrinking present and that of the ever-expanding present depend upon a confusion between instants and intervals. Provided we keep that distinction firmly in mind, we shall be able to understand what Plotinus, Augustine and Boethius were driving at, and see that it does not show that God is timeless, though it does help characterise God's experience of time in contrast to ours, and thus give us our idea of eternity.

Time and Religion

```
Past                    Present Year                    Future
-------------- * * * * * * * * * * * * *+ + + + + + + + + + + + + + +

Past                    Present Term                    Future
-------------- * * * * * * * * * * * * + + + + + + + + + + + + + +

Past                    Present Week                    Future
---------------- * * * * * * * * * + + + + + + + + + + + + + + +

Past                    Present Day                     Future
------------------ * * * * * * * * + + + + + + + + + + + + + + +

Past                    Present Hour                    Future
-------------------- * * * * * * + + + + + + + + + + + + + + + +

Past                    Present Minute                  Future
---------------------- * * * * + + + + + + + + + + + + + + + + +

Past                    Present Second                  Future
------------------------ * * + + + + + + + + + + + + + + + + + +

Past                    Present? Limit?                 Future
------------------------ | + + + + + + + + + + + + + + + + + + +

                     No Present Interval at all

ergo

                          No Time at all
```

FIGURE 6. Augustine's ever-shrinking present.

Eternity

Philosophers are much given to 'egotheism', and are very ready to take a God's-eye view of the universe: as I survey the whole of reality, the Big Bang and the Big Crunch are within the ambit of my thought, but their being present to my mind, and within the present aeon (note the two different senses of 'present'), does not preclude their being, in the one case before the present instant and therefore past, and in the other case after the present instant and therefore future. God's time differs from our time in the reach of the present *interval*. The time of our mortal life is brief. Most of the past

> Is **Plato** right to think of the philosopher as the Spectator of All Time?
>
> We can take up different temporal standpoints:
>
> When Darwin **went** on the Beagle, he **had already won** a good reputation in Cambridge.
>
> **Past**   **Present**   **Future**
> • • • • • • • • • • • •B• • • • • • • • |▷ ▷ ▷ ▷ ▷ ▷ ▷ ▷ ▷ ▷ ▷ ▷ ▷ ▷ ▷ ▷ ▷
> • • • • •C• • • • • |
>
> When Darwin **went** on the *Beagle*, he **was thinking** more about birds than girls.
>
> **Past**   **Present**   **Future**
> • • • • • • • • • • • •B• • • • • • • • |▷ ▷ ▷ ▷ ▷ ▷ ▷ ▷ ▷ ▷ ▷ ▷ ▷ ▷ ▷ ▷ ▷
> • • • • • • • • ✶ ✶ ✶ |✶ ✶ ✶ ▷ ▷ ▷ ▷
> • • • • • • • • • bbbbbbb▷ ▷ ▷ ▷
>
> Our thoughts are not restricted in the way physical causality is. I can adopt any temporal standpoint, much as I can imagine myself in your shoes, and see things from your point of view. We can think about the Big Bang or wonder what it will be like at the coming of the Big Crunch. So from a first-personal point of view, all time is conceptually present.
>
>   < < < < < < < <Present> > > > > > > >
> ✶✶✶|✶✶✶

FIGURE 7. The spectator of all time.

happened long before we were born, most of the future will happen long after we are dead. Not so with God. No past time is beyond His ken, no future time beyond His concern. His present interval embraces all our pasts and all our futures. This is the sense of Augustine's

> Thou precedest all past times by the eminence of thine ever-present eternity, and thou goest beyond all future times,

(though the mathematician in me wants to niggle and rephrase it in the delta-epsilon notation, and say that for every past date there is an antecedent one within God's ever-present eternity, but I will spare you my nigglomania) (Figure 8).

We can see why Plotinus, Augustine and Boethius think that eternity is static. If God's present interval embraces the whole of time, it is in that sense

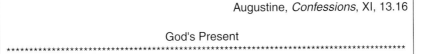

Praecedis omnia praeterita celsitudine semper praesentis aeternitatis, et superas omnia futura.
Thou precedest all past times by the eminence of thine ever-present eternity, and thou goest beyond all future times.

Augustine, *Confessions*, XI, 13.16

For any time in our past interval, there is a time in God's present interval which is before it,
and for any time in our future interval, there is a time in God's present interval which is after it.

FIGURE 8. Augustine's all-encompassing present.

always the same. But it does not follow that God's present *instant* must be always the same too. In contrasting eternity with time,

Nostrum nunc, quasi currens, tempus facit et sempiternum: divinum vero nunc, permanens neque movens sese atque consistens, aeternitatem facit,

Our present, as it were running, creates time and everlastingness: the divine present, remaining the same and not moving itself, creates eternity.

Boethius (*De Trinitate*, 4) is using the first *nunc* to refer to the present instant, the second to God's present interval. But time, whether ours or God's, whether just a finite period or everlasting time, is constituted by the moving present which divides future potentiality from the past, which cannot unhappen. Eternity is not untime, but all time, not timelessness, nor changelessness, but time not subject to the limitations of our mortal existence. Time teaches us our limitations (Figure 9).

We forget the past and fear the future, and often feel overwhelmed by the pressure of events; we just do not have time to get through the business of the day, and live in a perpetual state of having 'left undone the things we ought to have done'. But no part of the past is so remote as to be absent from the unforgetful memory of God, nor need He face the future with impotent foreboding; nor again is an infinite mind compelled to deal with only

> *Tempus sine aliqua mobili mutabilitate non est, in aeternitate autem nulla mutatio est.*
> There is no time without change, but in eternity there is no change.
>
> Augustine, City of God, XI, vi
>
> *In aeternitate stabilitas est, in tempore autem varietas: in aeternitate omnia stant, in temporalibus accedunt, alii succedunt.*
> In eternity there is permanence: in time variety. In eternity all things stand still; among temporal things some come, others follow.
>
> Augustine, Sermons, CXVII, 10(vii)
>
> *Aeternitas igitur est interminabilis vitae tota simul et perfecta possessio.*
> Eternity is complete, simultaneous and perfect possession of unending life.
>
> Boethius, Consolations of Philosophy, V, 6, ll. 9–11
>
> Time is the passage from future potentiality through present actuality to past immutability.
>
> Eternity is the contemplation of the whole of time with equanimity.

FIGURE 9. Eternity.

one thing at a time, and be unable to cope with the rush of modern life. The contrast is not between our time and God's timelessness, but between our limitation and God's lack of limitation. And that contrast is one that goes far beyond our finite intellectual capacities, to include our moral and personal deficiencies. It is a difference of values, not just a difference of scope (Figure 10).

Free will

We can see the significance of these evaluative issues if we turn to the further considerations that have inclined thinkers to adopt a timeless understanding of eternity. A timeless understanding of eternity avoids two awkward questions. In the first place it avoids our having to reconcile God's omniscience with human freedom. If God is outside time, He cannot know *in advance* what we are going to do – He just knows it timelessly. The God of Abraham, Isaac and Jacob could foretell, but His predictions were not infal-

> Eternity must imply everlasting time,
> **but**
> although the Greek words aidios (ἀίδιος) and aion (αἰών) naturally mean everlasting, there was a long tradition, from Philo onwards, of distinguishing eternity from everlastingness, or sempiternity as it was sometimes called. Eternity is not untime, but all time, not timelessness, but time not subject to the limitations of our mortal existence.
> The contrast is not a logical one, but an evaluative one:
> *Aeternitas igitur est* interminabilis vitae *tota simul et* perfecta possessio.
>
> *perfecta possessio*
> *interminabilis vitae*
> The vulnerability of God
> and
> the boringness of man

FIGURE 10. Is Eternity just Everlastingness?

lible, since He might warn us to mend our ways, and on occasion repented of the evil He had planned to visit upon us. But if God knows immutably and infallibly what we are going to do, it seems that we shall have no choice but to do it, and it is hard to blame us for what we were predestined to do. Philosophers may cavil and argue that to foreknow is not to cause, but to ordinary people the conclusion is inescapable: if God is omniscient, He knows what I am going to do; in which case, I shall not be able not to do it, and if it is a bad thing I am going to do, it is God's fault that it is done, since, being also omnipotent, He could have prevented me – if necessary by causing me to die. A road accident in Linz, a flu epidemic in Georgia, a mistaken murder by African freedom fighters, some fratricidal strife in Baghdad, and we should have been spared the horrors perpetrated by Hitler, Stalin, Idi Amin and Saddam Hussein. A traditional Christian response to the problem of evil has been what Plantinga calls 'The Free-Will Defence'. But the Free-Will Defence immediately calls into question the omniscience of God. If it is really up to me whether I go on the rampage or restrain my murderous instincts, listening politely to my pupils or colleagues as they commit fallacy after fallacy, God cannot be certain about what is going to happen at tomorrow's seminar: my restraint may finally break down, and I may dispatch the ill-deserving undergraduate with a carving knife.

Christian thinkers – Origen, Augustine, Milton – have sought to relieve

God of responsibility for our misdeeds by arguing that foreknowing is not the same as causing. That is true. But, as Jonathan Edwards, the first American to publish his thoughts on these matters, pointed out, it does not alter the case. God's foreknowledge may not cause me to sin, but, if it cannot be wrong, it cannot be the case that I shall not sin, so it cannot be up to me whether I do sin or not. At the time of decision there is no real possibility of my not sinning, and, if my sinning is a foregone conclusion, then, though it may be expedient to go through the motions of blaming me, it will not be just to do so. I cannot in all fairness be made to carry the can. God is the residual bearer of responsibility – and anyway, if He is omnipotent, He could have spared me from a fate worse than death by causing me to die. Boethius was sensitive to Jonathan Edwards's argument. Human freedom to decide future action precludes divine foreknowledge in the present and past. But to ascribe to God fallible opinions about what was yet to happen would be, so Boethius says, blasphemy (*Quod nefas judico*). The only way to preserve both divine infallibility and human freedom is to remove God from time altogether, so that His knowledge cannot be *fore*knowledge, and cannot be ascribed to Him now or at any time in the past, and so cannot be subsequently shown to be mistaken (Figure 11).

Such a freedom is purchased at a price. We protect God's infallibility, but deprive ourselves of our protector. God is no longer our comforter, our confidant, or our counsellor, if He is not here to help us when we seek His help, or to hear us when we want to open our hearts to Him, or ask His advice. The God of the philosophers is an Epicurean God, who leaves us alone because He is not interested in us – a better God than the capricious and sometimes malevolent deities of the pagan pantheon, but not a God to give solace to the soul or to support us amidst the manifold temptations of this troublesome life. But then why bother to believe in such a God, if it makes so little difference whether He exists or not? Rather than think of a personal deity that is unconcerned with human affairs, we do better as philosophers to think of an impersonal, because timeless, Ultimate Reality, which might also be the Ultimate explanation of everything, which we might well reverence and admire, but which it would be inappropriate to approach, to argue with, or seek absolution from. Plato used the neuter, τὸ θεῖον (*to theion*), of the Form of the Good, and later philosophers, more significantly since they were professing Christians, coined the term *ens realissimum*, again in the neuter, when talking of the Being to which Anselm had addressed his *Proslogion*.

I have drawn a stark contrast between the personal, and perhaps anthropomorphic, God of the Bible and the impersonal Ultimate Being of the philosophers, and in the end I believe that those are the only alternatives open to us. But the mainstream of Christian orthodoxy has hoped to be able to hold both these views in creative tension, and to find a way of embracing both in a higher synthesis. Augustine and Aquinas made use of the distinction between deciding to do something and actually doing it. It was some time ago that I

Q. How is it possible that there is a benevolent God when the world is full of misery and wickedness?
A. God has given man free will, and many men misuse it to do wrong.
Q. Is God then no longer omnipotent?
A. He still could intervene to stop men doing wrong, but that would be to circumscribe their freedom, so he chooses to limit the exercise of his omnipotence.
Q. So does he know the bad things I am going to do?
A. If he infallibly foreknew them, then it could not be the case that when the time came for me to act I might think better of it and change my mind. So I would not be free after all.
Q. Do you mean that God would have made you do them?
A. Not necessarily. It might have been my genes or my hormones: but it would not have been me, if I could not have changed my mind and repented of the evil I had in mind to do.
Q. So God is not omniscient?
A. It depends on what you mean by 'omniscient'.
Q. There are some things God does not know.
A. There are lots of things God does not know – that $2 + 2 = 5$, for example.
Q. Of course God does not know things that are false: but it was not false that you were going to misbehave, was it?
A. Maybe it was not true either .The Schoolmen had many debates about the truth-value of 'future contingents'.
Q. Too scholastic for me. Leave aside all questions of knowledge. Can God foretell the future? I can predict that there will be a Board of Philosophy Meeting next Monday. Is God ignorant of that?
A. God can foretell that too.
Q. And if someone on Monday morning thinks it is too nice a day to waste on a meeting, and rings around and persuades everyone to cancel the meeting, God will have foretold wrongly?
A. Yes.
Q. *Quod nefas judico.*

FIGURE 11. Free-will defence.

decided to write this chapter but only now that I am actually writing it. Although God's actions take place in time, His decision to do them might have been timeless. But that will not give us what we want. It might account for some major strategic moves in the story of our redemption, but cannot plausibly explain God's timely response to the situations in which we find ourselves and the petitions that we sometimes make. Unless we regard ourselves as mere puppets whose actions were determined long before we ever came into existence, we have to allow that our free decisions introduce new factors into the reckoning. A temporal God can reckon with them anew: a timeless being might be said to be timelessly minded to perform specified actions at specified times or in specified situations, but can never be more than a distant Deist Deity, remote and uninvolved in the doings and sufferings of particular people.

A God who spake by the prophets, or was incarnate in Jesus Christ, is a God not entirely outside time; if such a God did exist, His relations with time would be both simpler and more complicated than has been traditionally supposed. They would be simpler in that the Bible could be understood in a straightforward way without the need for re-interpreting all accounts of God's actions in the world and communications with men. Not only would we have a more intelligible account of God, but we should have a better view of man. Whereas the Preacher of Ecclesiastes in his atheistical despair held that there was nothing new under the sun, we should be able to believe that men could be genuinely creative, and that we had it in us to do something which was our own original contribution to the history of the world. However much we had messed things up hitherto, it would be possible to form a new mind μετάνοια (metanoia) and make a new start. We could leave behind us the things that were past and stretch out towards future aspirations, τὰ μὲν ὀπίσω ἐπιλανθανόμενος τοῖς δὲ ἔμπροσθεν ἐπεκτεινόμενος (ta men opiso epilanthanomenos tois de emprosthen epekteinomenos), seeing life not as a timeless entity or a meandering circle, but – a point recognised by some of the Eastern Fathers, notably Gregory of Nyssa – as a pilgrimage, a stretching out, an ἐπέκτασις (epektasis).

But if God is not a timeless being, we have to answer the difficult questions about foreknowledge and fallibility that traditional theologians were keen to avoid. Although the prophets were not always making predictions when they prophesied, they did on occasion predict, and the word of the Lord was vulnerable to subsequent falsification. Pharaoh might have not hardened his heart, and might have let God's people go with best wishes for their future happiness in a new land. The issue is not one of foreknowledge but forethinking: if man is free to make up his mind what he is going to do, then, if God cares at all about what we do, He will form opinions which may be falsified in the event: *quod nefas judico*. But is it really a blasphemy for a Christian to think that God might be disappointed in His expectations of what a particular person might do? The God of the philosophers is above that sort of thing. He needs to be infallible, or the philosophers will not think much of Him. But the gospel is foolishness to the Greeks. It speaks of a God who is vulnerable, and suffers a much worse fate than merely being wrong in some of His predictions.

The beginning of time

Augustine invoked the timelessness of God to sidestep the question of how God occupied His time before He created the world. We naturally speculate about origins. Even if time was not itself a thing made by God when He created the universe, it might only have come into existence then. Empiricists from Aristotle onwards have maintained that time involves change, and that where there is no change there is no time either. This has enabled them to avoid awkward questions about the beginning – What it was like in the beginning? What happened before the Big Bang? What was God doing before He created the world? – by saying that 'before the Big Bang' is a meaningless phrase. But 'before the Big Bang' is not a meaningless phrase: we can understand it quite well. For time does not imply change, but only the possibility of change – a point which, although evidently true, I found hard to prove to my satisfaction in the face of the verificationist ethos of our age. I did devise a rather thin, logic-chopping proof, using tense logic, but Sydney Schumacher produced a much better one, in which we are faced with a choice between having horrendously complicated laws of nature or allowing the existence of a temporal vacuum; that is to say an interval during which nothing whatsoever happened.

Time is not the measure of motion, but motion is characteristically used as the measure of interpersonal time. If we are to establish a way of measuring duration we all can use, we need to be able to recognise intervals that we can deem to be isochronous (of equal duration). We need processes subject to strong symmetry conditions, so that we can apply them to different durations on different occasions. Periodic processes are particularly suitable, and since, by some good fortune, there seems to be a natural rhythm in the universe, whereby all periodic processes keep in step, a large number of different clocks are available. We naturally project this requirement of measurability onto time itself, and think of it as homogeneous, and hence causally inert. But that is only the way we like to think about time, not the way it has to be in itself. If we do think of time as homogeneous and causally inert, we are vulnerable to Leibniz's challenge to Dr Clarke, of Caius College, to explain why God made the world when He did, and not a year earlier. Since, *ex hypothesi*, there are no antecedent circumstances to distinguish the two scenarios, there can, granted the translational symmetry of time, be no reason whatsoever why God should have chosen to create at the one time rather than the other. But to Leibniz' question we can return the Anti-Asinine Answer that when faced with perfect symmetry it is rational to break it arbitrarily in order to get one's teeth into something definite rather than remain suspended in hungry indecision. God could have chosen to set off the Big Bang when He did for no other reason than that He wanted to create sometime, and must at some time decide to get going. This is not something strange and mysterious. When I used to give the moral philosophy lectures in Oxford, I had to cover weakness of the will. My undergraduate audience were completely understanding when I directed their thoughts to being in bed and needing to get up, and how the knowledge that one ought to get up now failed to activate their limbs, but how eventually, for no particular reason, they finally did decide to get up. From the outside we cannot explain why one time rather than another, but from the inside we well appreciate the force of the answer 'I just decided to'. When we think about time from an impersonal point of view, we think of it as homogeneous, all times being alike: but, when we think of it from a first-personal point of view, we naturally differentiate between dates, the present being of pre-eminent concern, and different from the future, which in turn is different from the past. New Testament scholars sometimes distinguish καιρός (*kairos*) from χρόνος (*chronos*). Linguistically the distinction is open to attack, but conceptually the distinction is one that needs to be made (and was formulated long ago by F. M. Cornford in his *Microcosmographia Academica*), and articulates the thought that the time could be ripe for God to create the world without there being any chronological measure to measure out the years, or any reason chronologically speaking, why that date should be chosen in preference to any other.

Leibniz's challenge can be met. But if it is meaningful to ask questions about time before the Big Bang or, supposing the universe does come to an end, after the Last Day, we are exposed to the questions that embarrassed Augustine, 'What was God doing before He created the world?' and 'What will He do after the Day of Judgement?'. Some thinkers in the ancient world worried, in a very modern fashion, that He might be being idle. We can picture the Higher Education Funding Council sending a form to God for Him to fill in, to prove that He ought not to be made redundant, because He really is pulling His weight. The Future Projects column looks quite good – many co-authorships with people such as Isaiah, Deutero-Isaiah, Jeremiah, even Royalty such as David and Solomon, and the citations index is going to look very good indeed, but the Work Actually Completed and the Current Activities columns are blank, and there is no chance of even a two-star rating (Figure 12).

Higher Education Funding Council
Quality Assessment

Date: 4005 BC

Name: God

College(s): Trinity, Trinity Hall, Jesus, Christ's, Emmanuel, Corpus Christi

Degrees: (excluding Oxford and Cambridge MAs): None

Future Publications (list co-authors): Amos, King David, Ezekial, Ezra, Habakkuk, Haggai, Hosea, Isaiah (i), Isaiah (ii), Isaiah (iii), James, Jeremiah, Joel, John, Jude, Luke, Malachi, Mark, Matthew, Micah, Moses, Nahum, Obadiah, Peter, Paul, King Solomon, Zechariah, Zephaniah.

Future Projects: Create world, evolve man, select chosen people, advise prophets, go Myself to put matters right, provide continuing education and occasional inspiration, etc.

Current Activities (excluding reflection and cogitation): None

---

FIGURE 12. God's quality assessment form.

Conclusion

We need not pursue this line of enquiry further. The very arrogance of the questions, as I have phrased them, tells us all we need to know of the questioner and his culture, evidently unmindful of the psalmist's injunction 'Be still then, and know that I am God'. It projects on God our own limited resources. We are easily bored, and suppose that God might be likewise. Our concern becomes more personal when we move from thinking about God, before He had our world to worry about, to thinking about ourselves if we are fated to enjoy – or endure – unending existence in an after-life. The young Kipling, it is said, told his brother that, if he was good, he would go to heaven and play a harp sitting on a cloud. 'And if I am not?', enquired his younger brother. 'The alternative is even worse.' More modern fantasies run heaven and hell together. Lord Bruce-Gardyne, writing in the *Spectator* about his own impending death, told of a fisherman who found himself with a fishing rod by the crystal stream, itself filled with an abundance of fishable fish. But eventually even fishing palled, and he asked an attendant if he could do something else. 'No, Sir. Fishing is your line.' 'But this is intolerable; I might as well be in hell.' 'Where did you think you were, Sir?' Time is a severe test of value. Very little of what we do survives the test of time. Most of our life is frittered away on trivial pursuits that fail to be recognised as such only because they are transitory and are over and done with before we discover their vacuity. But if we had all eternity to occupy, the emptiness of our endeavours could not be evaded by passing on to other enterprises equally empty. In this life we can divert our thoughts from our true situation, and by many *divertissements* seek to kill time until finally time kills us; but when we have all the time in the world, many things we prize now because life is short will no longer have any rarity value, and will cease to be precious in our long-distance eyes. When everything that is possible has become actual in the fullness of time and there is nothing new under the sun, we fear that we shall find, like the Preacher, that all is vanity, and the whole of existence a vanity of vanities, as we eat out our hearts by the quiet waters of futility.

I cannot answer that charge or assuage that fear. My values are as circumscribed as yours, and I have no warrant to affirm that they will not be eroded by time. But it is not clear that all values must be eroded, or that some Second Law of Spiritual Dynamics necessitates a gradual run-down into eternal tedium. It could be so, but it does not have to be so. And indeed we have some grounds for hope. The view of the universe that associates continual degrading with the passage of time is a closed view, and there are reasons for holding that any closed view is conceptually inadequate. Theism, which claims that the fundamental reality is a personal being and the fundamental explanation a personal explanation, is often accused of being anthropomorphic. But though human beings have feet of clay, they also have infinite aspirations, aspirations they are not altogether unable to realise. There is an infinite diversity about humanity that might give grounds for infinite occupation and enjoyment. And if God is, as the Christians aver, a God with a human face, it could be that infinite existence was something to enjoy rather than endure.

That is a hope. It is only a hope. We cannot be sure. All I have attempted to show is that, contrary to much theological teaching, religion does not need to make out that God is timeless, or that time is in some fundamental sense unreal. A vulnerable God can be temporal, exposed to the future ill will of autonomous agents. If God has created us free, then only in some eschatological Kingdom of Ends will He not be hurt by our imperfect choices. Perhaps that is too remote a possibility for us to look for. We can only look forward a little way, limited by death. Our search for the ultimate truth or the end of existence is necessarily a search without certitude, and within the compass of what we know, the atheist can only hope, and the Christian pray, that death when it comes will come no longer as an enemy, but not yet as a friend.

FURTHER READING

Kirwan, C., *Augustine*, Chapters 7, 8 and 9, London: Routledge, 1989, paperback 1991.

Lucas, J. R., *A Treatise on Time and Space*, London: Methuen, 1973, especially §§55 and 56.

Lucas, J. R., *The Future*, Oxford: Blackwell, 1989.

Sorabji, R., *Time, Creation & The Continuum*, London: Duckworth, 1983.

Notes on Contributors

Dame Gillian Beer is King Edward VII Professor of English Literature at the University of Cambridge and President of Clare Hall. Among her books are *Darwin's Plots* (2nd edition, 2000), *Open Fields: Science in Cultural Encounter* (1996) and *Virginia Woolf: The Common Ground* (1996).

David Crystal is a writer, lecturer and broadcaster on language and linguistics, reference books editor, and honorary professor of linguistics at the University of Wales, Bangor. He is editor of a family of general encyclopaedias, and his recent authorial writing includes *Words on Words: Quotations about Language and Languages* (Penguin), and *Language and the Internet* (Cambridge University Press).

Christopher Isham has been professor of theoretical physics at Imperial College London since 1982. His research during the last 25 years has concentrated mainly on mathematical and foundational issues in quantum gravity and quantum theory – in particular, the role that time plays in these areas. His main interests outside theoretical physics and mathematics are philosophy, the work of C. G. Jung, theology, and classical music – especially singing Gregorian plainsong.

Charalambos P. Kyriacou is Professor of Behavioural Genetics at the Department of Genetics in Leicester University. His degrees are in Psychology and Genetics, and his interests are in the molecular genetic analysis of behaviour, using the fruitfly *Drosophila* as the model system. He has been studying sexual behaviour and circadian rhythms in the fly since 1973.

Notes on Contributors

J. R. Lucas is the author of *A Treatise on Time and Space; Space, Time and Causality; The Future* and (jointly) *Spacetime and Electromagnetism*; he has contributed to *The Arguments of Time* (Oxford University Press) and *Time, Creation and World-Order* (Aarhus University Press). He was Fellow and Tutor of Merton College, Oxford. He was President of the British Society for the Philosophy of Science. He is a Fellow of the British Academy. His website is http://users.ox.ac.uk/~jrlucas

D. H. Mellor is Emeritus Professor of Philosophy at Cambridge University, and a Fellow of Darwin College and of the British Academy. His main philosophical interests are in metaphysics, especially the nature of time, causation and chance.

Katinka Ridderbos is a Research Fellow at Darwin College, Cambridge. Her main research interests are the conceptual foundations of statistical mechanics, and the problem of the direction of time.

Konstantina Savvidou currently works at Imperial College, London. Her PhD thesis in theoretical physics put forward the idea of the appearance of two distinct modes of perceiving the concept of time and its application to consistent histories theory. Her major research interest remains the study of the problem of time in quantum gravity, and in particular the application to this problem of her novel ideas about time. She is a strong believer in an interdisciplinary study of the Cosmos, with ideas from both the sciences and the arts contributing to our understanding of Nature.

Romila Thapar, a historian of early Indian history, is currently Emeritus Professor of History at the Jawaharlal Nehru University in New Delhi. She has been a Visiting Professor at Cornell University and the University of Pennsylvania and has lectured at the Collège de France. In 1983 she was General President of the Indian History Congress. She is an Honorary Fellow of the School of Oriental and African Studies and of Lady Margaret Hall at Oxford, and was elected as a Corresponding Fellow of the British Academy. Among her publications are: *Asoka and the Decline of the Mauryas*; *A History of India*, vol. I, *From Lineage to State*; and *Time as a Metaphor of History*.

Alan Wing is Professor of Human Movement in the School of Psychology at the University of Birmingham and director of the Sensory Motor Neuroscience Research Group in the new University Behavioural Brain Sciences Centre. In addition to his MRC funded research on timing he is interested in anticipatory mechanisms of posture and movement and also in active touch.

Acknowledgements

COVER ILLUSTRATION: Copyright Digital Vision Ltd.

CHAPTER 1

C.J.I. and K.N.S. are both grateful to Jeremy Butterfield for his very helpful comments on a preliminary version of this chapter. K.N.S. would like to especially thank Father Stamatios Skliris for very enlightening discussions through several years, and C. Anastopoulos for his help in collecting parts of the material used for this work. K.N.S. was supported by a gift from the Jesse Phillips Foundation.

FIGURE 3: Courtesy of John Latham.

FIGURE 4: From Cumont, F., *The Mysteries of Mithra*, New York: Dover Publications, 1956, with permission of Dover Publications, Inc.

CHAPTER 2

PLATE I: From the Calico Museum, Ahmedabad, Gujerat, India; courtesy Sarabhai Foundation, Shahibag, Ahmedabad 380 004.

FIGURE 2: From Hultzsch, E., *Inscriptions of Asoka, Corpus Inscriptionum Indicarum*, vol. 1, Oxford: Clarendon Press, 1925. Copyright 1925 Clarendon Press, by permission of Oxford University Press.

FIGURE 3: From Sircar, D. C., *Indian Epigraphy*, plate XXV, Delhi: Molital Banarsidass, 1965. Copyright Archaeological Survey of India.

FIGURE 4: From Archaeological Survey of India: *Epigraphica Indica*, vol. 6, Calcutta. Copyright Archaeological Survey of India.

CHAPTER 4

C.P.K. thanks Ben Collins for drawing the prototypes for Figures 3 and 4.

CHAPTER 5

FIGURE 1: Johansson, R. S. and Westling, G., 'Programmed and triggered

Acknowledgements

actions to rapid load ch anges during precision grip', *Experimental Brain Research*, **71** (1988), 72–86. Copyright 1988 Springer-Verlag.

FIGURE 3b: From Wing, A., 'The long and the short of timing in response sequences', in *Tutorials in Motor Behaviour*, ed. G. E. Stelmach and J. Requin, pp. 469–486. Amsterdam: North-Holland, 1980.

FIGURE 4c: From Sergent, V., Hellige, J. B., and Cherry, B., 'Effects of responding hand and concurrent verbal processing on time-keeping and motor-implementation processes', *Brain and Cognition*, **23** (1993), 243–262, by permission of Academic Press.

PLATE I: From Rao, S. M., Harrington, D. L., Haaland, K. Y., Bobholz, J. A., Cox, R. W. and Binder, J. R., 'Distributed neural systems underlying the timing of movements', *Journal of Neuroscience*, **17** (1997), 5528–5535. Copyright 1997 by the Society for Neuroscience.

CHAPTER 6

FIGURE 3: From *The Independent*, Sports Section, 20 September 1999, with permission from Independent Newspapers (UK) Ltd.

FIGURE 4: From Crystal, D. (ed.), *The Cambridge Factfinder*, 4th edition, Cambridge: Cambridge University Press, 2000, with permission of Cambridge University Press.

FIGURE 5: From Crystal, D., *The Cambridge Encyclopedia of Language*, 2nd edition, Cambridge: Cambridge University Press, 1997, with permission of Cambridge University Press.

CHAPTER 7

'The Day I Walked in Wet Cement': From Cambridge Young Writers, *I'm Telling You!*, Cambridge: Cambridge University Press, 2000. © Cambridge University Press and Bryan Clark.

FIGURE 2: From Clark, B., 'The Day I Walked in Wet Cement', in Cambridge Young Writers, *I'm Telling You!*, Cambridge: Cambridge University Press, 2000, with permission of Lizzie Finlay.

FIGURE 3: Copyright DACS 2001.

Index

Page numbers in **bold** refer to figures

Aboriginal languages 119, 121
Adams, Douglas 105
Anselm 159
Aristotle 8-9, 161
Augustine 1, 7-9, 144, 150-63, **153**, **155**, **156**
Austen, Jane 134, 136

beginning 2, 8, 28, 32, 36, 37, 47, **71**, 73, 81, 124
Benjamin, Walter 132, 141
Big Bang 152-3, **154**, 161-2
Big Crunch 152-3, **154**
biographies 40-2, 44
Boethius 147, **148**, 152, 154-5, 158
bradykinesia 100-1
brain 74, 77-8
  control of movement by 76, 100, **101**
  damage 4, 100
  imaging 4, 100, 102
Bruce-Gardyne, Lord 164
Buddhist/Buddha 30, 32-3, 38-9, **39**

Calvino, Italo 137-8
Cantor 151
Carroll, Lewis 131
Carson, Ciaran 127-9, 132
cerebellum 100, 102
change 2-3, 5, 9, 13, 19-20, 28-30, 32, 36, 42, 46, 70, 76, 80-1, 85-8 **87**, **89**, 99, 101, 106, 117, 123, 129, 133, 135, 140, 141-7, **144**, **146**, **156**, **159**, 161
Christianity 10 , 33, 143-4, 157-9, 161, 165
  Greek Orthodox Church 8
chronicles 27, 38-9, 41-2, 44-5
circadian 3, 65-70, **72**, 73, 75-81, 83

Clark, Brian 130
Clock (gene) 74, 78-9
CLOCK (protein) 75-8
clocks 2-4, 7, 23, 48-9, 66-70, 73, 75-83, **81**, 122-3, 162
  external 47, 67
  internal 1, 2, 5, 47, 67, 94
Cobbett, William 107-8
conditional tense 105, 107
constellations 29
creation 4-5, 10, 24, 30-1, 37-8, 42, 44-5, 64, 120, 144, 146, 155, 161-3, 165
CRYPTOCHROME (protein) 76-9
culture and language 4, 115-19
cycle 3, 27-8, 30-3, 37, 39, 42, 65, 68-70, **72**, 73-9, 81, 83, 97, 99, 117
cycle (gene) 74
CYCLE (protein) 74-8
cyclic time 1, 2, 5, 27-8, 30-1, 33, 35, 42-4

Darwin, Charles 150, 152, **154**
delay 68, 74-7, 133-4
Demiurge 147
differential calculus 13, 18, 26
differential geometry 19, 26
discernible 17
doubletime (gene) 76, 78
Dr Who 46, 48-9
Drosophila 69, 71, **72**, 75
duration 17-19
dynasties 35-9, **39**, 42-4

Edwards, Jonathan 158
egotheism **151**, 153
Einstein, Albert 49, 123, 147

Einstein's field equations 21–2
electromagnetism 148–9
electrons 53–7, **54**, **57**
Eliade, Mircea 28
English 106–9, **106**, 112–19, 120–2, 123–4, 127–8, 129, 135–6
ens realissimum 158–9
eras 33, 37–8, 40, 44
eternity 8–10, 28, 36, 144, 152–6, **155**, **156**, **157**, 164
ether 149
ever-expanding present 151–2
ever-shrinking present 150, 152–3

Fielding, Henry 131
fitness 80–1
flood 35–6
foreknowledge **144**, 145, 158, 161
free-running rhythms 68, 75, 82
free will 5, 145, 156–7, **159**, 160
future 8–10, 13, 45, 47–8, 59, 63–4, 77, 103, 105, 110, 112–16, 118–19, 121–2, **123**, 124, 126–7, 132–41, 148, 150–5, **152**, **153**, **154**, **155**, **156**, 158, **159**, 160–1, 162, 165
futures 4, 127, 133–6, 138, 140–1

gene 70, 73–82
genealogies 35–7, 40, 42–4
General Theory of Relativity 7, 9–10, 21–3, 150
God 8, 24–5, 34–5, 146, **151**, 153–5, 157–8, **157**, 159–60, **159**, 162, 164
  and infallibility 156–7, 158, 161
  and timelessness 5, 143–5, **144**, 147, 152, 156, 160–1, 165
  outside time 147, **148**, 156, 158, 160
grammar 4, 105–9, 115–16, 119, 121
Grass, Günter 138–40
Grimm, brothers 138–40

Hall, Edward T. 117–18
Hardy, Thomas 130
history 2, 10, **15**, 16, 20–2, 25, 27–8, 31, 33, 36–41, 43–5, 126, 138–9, 143–4, 160
Hopi 4, 115, 116

Indologists 40
industrial accidents 67
inflections 115–16, 119
inscriptions 37, **39**, 40, **41**, 42, **43**, 44
instant 5, 18, 59, 145, 150–3, **152**, 155
interaction, presence and 3, 51–2, 63
interval 5, 71, 86, 88–91, **92**, 93–4, **93**, **95**, 96–7, **98**, 99, **99**, 102–3, 133, 145, 150–5, **152**, **153**, **155**, 161

Japanese 115–16
jet-lag 3, 66–8, 83
Joyce, James 126

Kafka, Franz 135
Kaliyuga 31–3, 35, 37, 42–3
Kalkin 33, **34**
kalpa 30–1
Kierkegaard, Soren 131
Kipling 164
kshatriya 33, 35–7

Latham, John 16, **17**
Latin 106, **107**, 108, 122, 132
Leibniz 7, 162
light 3, 49, 51, 55, 66, 68, 75–7, 79, 81, 137
linear time 1–2, 5, 27–8, 33, 35–8, 40–2, 44–5, 116
Lorentz 148–9
love songs 65, 71, **72**, 73

Mach 7
Maximus the Confessor 8–9
measurement 3, 7–9, 19–21, 29–31, 37, 45, 47, 86, **87**, 88, 96, 120, 149, 161–3
memory 94, **95**, 102–3, 134, 155
messenger RNA (mRNA) 74–9
Michelson–Morley experiment 149
Milton 158
minds **151**, 161
Minkowski 147
motor cortex 100, **101**
motor implementation delays 85, 88, 90–1, **93**, 94, **95**, 96–7, 100, 102–3
Murray, Lindley 106, 108, 112

Index

negative feedback 74, 79
neurological disorders 4, 100–1, 103
Newton 11, 20, 116, 147, 149–50
Newtonian mechanics 148–9
Newtonian physics 6–7, 9, 13, 21–2
Newton's laws of motion 13, 18
nucleus 74–8

Old Testament 144, 149
omniscience **144**, 149, 156–7, 159
Orientalists 27, 35
Origen 157

pacemakers 77–8, **95**
passion 133, 138
past 2–5, 7–10, 28–9, 35–6, 42–5, 51, 59–60, 62, 64, 105, 109–10, 113–17, 118–19, 121, **123**, 124, 127, 132–3, 136, 138, 140–1, 148, 150–5, **152**, **153**, **154**, **155**, **156**, 158, 160, 162
perfection **144**, 145–6
period 66, 68–70, **71**, **72**, 76, 81–2, **81**
period (gene) 69–70
PERIOD (protein) 70, **71**, **72**, 73–83
phosphorylation 76
planets 30
Plantinga, A. 157
Plato 5, 145–6, 151, **154**, 158
Plotinus 152, 154
polyrhythm 99–100
positrons *see* electrons
Potawatomi 115
present 5, 8–10, 32, 38, 42, 45, 110, 114, 118, 121, **123**, 129, 132, 135, 140–1, 147–8, **148**, 150–5, **152**, **153**, **154**, **155**, **156**, 158, 162
principle of excluded middle 19
principle of relativity 148–50
process philosophy 16–17, 19, 23
propositional calculus 12–13
protein 74–7, 79, 81, 83
Purana 33–5, 37, 39, 42–4
   Vishnu 33, 35, 37, 43
   Itihasa 36

quantum gravity 2, 13, 22–4, 26
   canonical quantisation approach to 23

quantum mechanics 150
quantum theory 10, 15, 19–24, 26
   consistent histories approach to 20

Rajatarangini 41
reader 5, 126–9, 131–40
real numbers 9–10, 13, 18–19, 25–6, 151
resetting clocks 3, 68, 75, 80, 83
resonance 80–3, **81**
rhythm 3–4, 65–6, 68–9, **71**, **72**, 73, 75, 78–9, 83, 97, **98**, 99–100, 102–3, 162
Richardson, Samuel 131, 135

Second Law of Spiritual Dynamics 165
Shakespeare 121
shall vs. will 112–13
shudra 33
sign language 123–4, **123**
sleep 65–6, 70, 73, 76, 122, 131
space 5–8, 11, 16–17, 21–2, 45–6, 48, 51–3, 59, 90, 124, 133, 140, 145, 147–50, **148**
spacetime 21–3, 50, 147, 151
   curved 21–2
   flat 21
Special Theory of Relativity 6–7, 9, 21, 49, 148–9
state space 10–16
Steiner, George 116
Stevens, L.T. 87–8, **89**, 90, 96
storyteller 4–5, 127–31, 132–3, 137
storytime 126
synchronisation 86, 90, 96–7, 103
system state 9, 11, 13–15

TARDIS 48–52, 59–63, 60
temperature compensation 69, 82
temporal logic 7, 9–10, 13–16, 19, 20–1, 24–5
tense 127, 135–6, 150, 152
   future 4, 36, 105–8, **107**, 112–13, 115, 127, 138
   imperfect 106, **107**
   past 4, 106–8, 113, 115, 128, 132, 135
   perfect 105–6, **107**, 108, 127, 138
   pluperfect 106–7, **107**
   present 4, 106, **107**, 108–10, **109**, **111**, 113, 115, 131–2, 135

Index

Tertullian 143, **144**
theism 146-7, 165
time
  absolute view of 6-7, 9
  and duration 73, 88, 97, **98**, **99**, **152**, 162
  and minds 150-2, **151**
  as an ordering parameter 9-10, 25
  as the parameter of evolution 9
  beginning of 26, 161-3
  creation of 147
  cyclic ideas of 10
  direction of 52-8, **56**, **57**
    causation and 55, 61
    divergence of light and 55
    expansion of the universe and 55
    increasing entropy and 55
  linear ideas of 10
  machines 47-8, 50-1, 58-62
  moment of 2, 9, 11, 14, 17-18, 22, 40
  of becoming 2, 9, 15-19, 24-5
  of being 2, 9, 11, 15-19, 21, 24-5
  origin of 24
  passing 3, 46-8, 50, 53, 55, 119-21, 124, 164
    rate 46-7, 49
  relational view of 6, 23
  see also cyclic time
timekeeper 88, 90-1, **92**, 93-4, **93**, **95**, 96-7, 99-103, **99**
TIMELESS (protein) 75-80
timelessness 110, 114, 131, 155, **157**, 158-60
time-line 4, 106, 109, 113, 115
timing
  of movement 3-4, 86-91, 103
  predictive 85-6

two-stage model of 88, 90-1, **93**, 96-7, **98**, 99-100, 103
  variability in 88-91, 93, **93**, 94, 96-7, 101-3
time travel 1-3, 5, 46, 60-3, 105
  backward 3-4, 48, 50-5, 57-64
    objections to 50, 58, 62
  forward 2, 47-50, 52, 59
    cryonics and 49
    hibernation and 49
    relativity and 49
Tiv 116-17
travelling
  arriving and 48, 50-2, 54, 58-61, 63-4
  in time, problems of 57-64

ultimate explanation 146, 158
unripe time, principle of 162

Varahamihira 29
verb 106-7, 112-16, 121-2, 123

Warner, Sylvia Townsend 126
Whitehead, Alfred N. 17-19
will vs. shall *see* shall vs. will
Wing-Kristofferson model *see* timing, two-stage model of
Woolf, Virginia 133, 140
world lines 52, **53**, 54-7, **54**, 59, 61
writer 127-35

yuga 30-2

Zervan 24, **25**